Lecture Notes in Electrical Engineering

Volume 306

For further volumes:
http://www.springer.com/series/7818

About this Series

"Lecture Notes in Electrical Engineering (LNEE)" is a book series which reports the latest research and developments in Electrical Engineering, namely:

- Communication, Networks, and Information Theory
- Computer Engineering
- Signal, Image, Speech and Information Processing
- Circuits and Systems
- Bioengineering

LNEE publishes authored monographs and contributed volumes which present cutting edge research information as well as new perspectives on classical fields, while maintaining Springer's high standards of academic excellence. Also considered for publication are lecture materials, proceedings, and other related materials of exceptionally high quality and interest. The subject matter should be original and timely, reporting the latest research and developments in all areas of electrical engineering.

The audience for the books in LNEE consists of advanced level students, researchers, and industry professionals working at the forefront of their fields. Much like Springer's other Lecture Notes series, LNEE will be distributed through Springer's print and electronic publishing channels.

Rainer Tutsch · Young-June Cho
Wei-Chih Wang · Hyungsuck Cho
Editors

Progress in Optomechatronic Technologies

Proceedings of the 2013 International Symposium on Optomechatronic Technologies, Oct 28–30, 2013, Jeju Island, Korea

Springer

Editors
Rainer Tutsch
Technical University Braunschweig
Braunschweig
Germany

Wei-Chih Wang
University of Washington
Seattle
USA

Young-June Cho
Korea Institute of Industrial Technology
Cheonan-si
Republic of Korea (South Korea)

Hyungsuck Cho
Korea Advanced Institute of Science
and Technology
Daejeon
Republic of Korea (South Korea)

ISSN 1876-1100 ISSN 1876-1119 (electronic)
ISBN 978-3-319-05710-1 ISBN 978-3-319-05711-8 (eBook)
DOI 10.1007/978-3-319-05711-8
Springer Cham Heidelberg New York Dordrecht London

Library of Congress Control Number: 2014936843

Printed on acid-free paper

Springer is part of Springer Science+Business Media (www.springer.com)

Preface

Optomechatronics, as fusion of optical and mechatronic engineering, has played a key role in developing innovative products such as high precision instruments, defense, photonic systems, measurements, diagnostics, semiconductors, and so on. Furthermore, optomechatronic technologies have greatly contributed to the state of the art of industries in optics design, manufacturing, optical imaging, metrology, and other applications. By using the optomechatronic technologies, we are able to create innovative technologies, generate a new value, and thus provide sophisticated systems consisting of optical, mechanical, and electronical technologies. This book aims at covering a number of optomechatronics advantages and solutions. It includes 20 contributions that feature laser and fiber optics, nitride semiconductors, LIDAR technology, machine vision, optical imaging, micro-optoelectro-mechanical systems, optical metrology, optical-based sensors and actuators, optomechatronics for microscopes, optical pattern and fiber, optomechatronics for biomedical applications, optomechatronics for manufacturing applications, robotics for micro and nanoscales, and other applications. As revised and extended versions, the contributed articles are selected from the "Proceedings of the 2013 International Symposium on Optomechatronic Technologies held on Oct 28–30, 2013 in Jeju Island, Korea." This edition is published in original, peer reviewed contributions covering the topic from research results and theories to verifications.

The 20 articles compiled in this book are representative of the broad range of methods and applications of optomechatronics. To help readers understand this book, we classify the 20 articles into 4 parts in area order.

The first part of six articles describes the current status and future prospects of optical imaging technology and application.

1. Integration of Image and ID-POS in ISZOT for Behavior Analysis of Shoppers
2. Low-Cost IR Visualizer Based on a Rotating Phosphor Screen for Accurate Beam Analysis
3. Evaluation of Heteroepitaxially Grown Semipolar {20-21} GaN on Patterned Sapphire Substrate
4. Development of a Low-Noise Three-Dimensional Imaging LIDAR System Using Two 1 × 8 Geiger-Mode Avalanche Photodiode Arrays

5. Quick Light-Mixing for Image Inspection Using Simplex Search and Robust Parameter Design
6. Dense 3D Reconstruction in Multi-camera Systems.

The second part of 3 articles describes optical metrology including measurement systems of microsphere, 3D microscope, and large surface.

7. Optimization of Coupling Condition in Distance between the Sphere and the Tapered Fiber for Diameter Measurement of Microsphere by Using WGM Resonance
8. UV Shadow Moiré for Specular and Large Surface Shape Measurement
9. Development of Measurement System for 3D Microscope

In the third part, there are 4 articles using optical-based mechanical systems including sensors and actuators.

10. Design and Fabrication of a Prototype Actuator for Fourier Transform Interferometry
11. Peak Search Method for the Optical Spectrum Signals
12. Magnetostriction-Induced Fiberoptic Metal Profile Detector
13. Liquid Viscosity Sensing Using Nonlinear Higher Order Harmonic Excitation in Fiberoptic Vibrating Sensor.

Lastly, the fourth part of 7 articles deals with optomechatronics applications in biomedical, nonhuman surgical assistants, manufacturing, and robotics.

14. Simulation and Optimization of Nanoparticle Patterned Substrates for SERS Effect
15. An Image-Based Coordinate Tracking System Using a Focal Optics for Surgical Navigation
16. Feasibility Study of a Functional Near Infrared Spectroscopy as a Brain Optical Imaging Modality for Rehabilitation Medicine
17. Real-Time Monitoring System for Industrial Motion and Optical Micro Vibration Detection
18. Thermal Manipulation Utilizing Micro-cantilever Probe in Scanning Electron Microscopy
19. Sound-Source Tracking and Obstacle Avoidance System for the Mobile Robot
20. Object Tracking Robot Using Data Combination of Ultrasonic Sensor and Servomotor.

We hope to help readers gain useful information from the articles and create an innovative and novel concept or theory. Thank you.

Rainer Tutsch
Young-June Cho
Wei-Chih Wang
Hyungsuck Cho

Contents

Part I
Current Status and Future Prospects of Optical Imaging Technology and Application

Chapter 1
Integration of Image and ID-POS in ISZOT for Behavior Analysis of Shoppers

Toshiki Etchuya, Hiroyuki Nara, Shun'ichi Kaneko, Yuan Li, Masanori Miyoshi, Hironobu Fujiyoshi and Kotomi Shishido

Abstract Recently the tendency of security oneself is growing because of increasing crimes. So a number of security cameras and the place setting the cameras are increasing, but those cameras are not used effectively. They are usually used for arrest a criminal after crime happens. The research analyzing behavior of shoppers by using a security camera have been studied. There are some methods, using the behavior of part of body, using the trajectories. In this paper, use the trajectories and analyze behavior of shoppers and introduce a method of calculate similarities and clustering them.

Keywords Shopper's trajectory • Zone • Point of sales • POS2ZT

1.1 Introduction

In recent years, in-store merchandising (ISM), which analyze shoppers' behavior in stores, has been attracting attention. ISM is analysis of shoppers' behavior in store to understand the effective way of advertisement or efficient product arrangement. It lead average sales figure per shopper bigger.

Shoppers' behavior in stores is defined by six actions (come into a store, pass a store shelf, stop in front of a store shelf, pick up, take into basket and buy) in ISM.

T. Etchuya (✉) · H. Nara · S. Kaneko
Hokkaido University, Sapporo-shi, Hokkaido, Japan
e-mail: etchuya@ssc.ssi.ist.hokudai.ac.jp

Y. Li · M. Miyoshi
Hitachi Research Laboratory, Hitachi Ltd, Hitachi-shi, Ibaraki, Japan

H. Fujiyoshi
Chubu University, Kasugai-shi, Aichi, Japan

K. Shishido
Consumers Co-operative, Sapporo-shi, Hokkaido, Japan

R. Tutsch et al. (eds.), *Progress in Optomechatronic Technologies*,
Lecture Notes in Electrical Engineering 306, DOI: 10.1007/978-3-319-05711-8_1,

Ordinary, point of sales (POS) system is used for ISM. POS system is resister system of sales merchandises in casher. POS record purchased merchandises and time. It can become understandable what is best-selling and combined buying merchandises. In addition to POS, TESCO in America and CO-OP Sapporo in Japan analyze shoppers' behavior with personal data. It's called ID-POS and it has age, sex, address, and some other personal data. By using ID-POS, they can understand shoppers' behavior deeper than that only by POS. POS or ID-POS give information for analysis of shoppers' behavior 'buy'.

However they can't record order of them or behavior of shoppers in the store. Also we can't know merchandises which shopper considered and didn't buy or behavior of shoppers who bought nothing. Some researchers investigate the method to get trajectories using RFID and analyze them [1]. However, initial cost is expensive.

We propose the method to use a security camera. Most of the stores have some security cameras because of an increase of senses for safety and decreasing a price of a security camera. However, most of the videos taken by security cameras are not used. Previously, analysts check all videos or traces after shoppers in a store for behavior analysis. It takes many times and cost. If shoppers are clustered by similar behavior, analysis data decrease and we can easily analyze shoppers' behavior in a store. As the similar merchandises locate closely, the shoppers who have the same purpose became the same trajectory. If we cluster shoppers by the trajectory, we can cluster by purposes of action. In this paper, we introduce a method for human detection and extracting trajectories from security cameras, clustering shoppers by their trajectories. In addition, we propose a method to connect POS and shoppers trajectories by ISZOT. A person who comes to the store is called as "shopper" in this paper. It is defined as ISZOT(Image based sensing system for analyzing shopper's zone trajectory) that get the shoppers' trajectory and analyze shoppers' behavior using shoppers trajectories. ISZOT has 4 parts, human detection, tracking, similarity calculation, clustering. In this paper, explain detail of similarity calculation and clustering.

1.2 Integration of Image and ID-POS

In this section, we explain three methods, human detection method using SWT (Smart Window Transform), similarity calculation method of Zone-based trajectories and Integration of image and ID-POS. Figure 1.1 shows the system flow. Transform position-based trajectory to zone-based trajectory and calculate similarities among all sets of two trajectories. Make clusters using hierarchal clustering using similarities (Fig. 1.2).

1.2.1 Human Detection Method Using SWT (Smart Window Transform)

There are many methods for human detection. Most detection methods of recent years have used object classification methods that apply statistical training methods

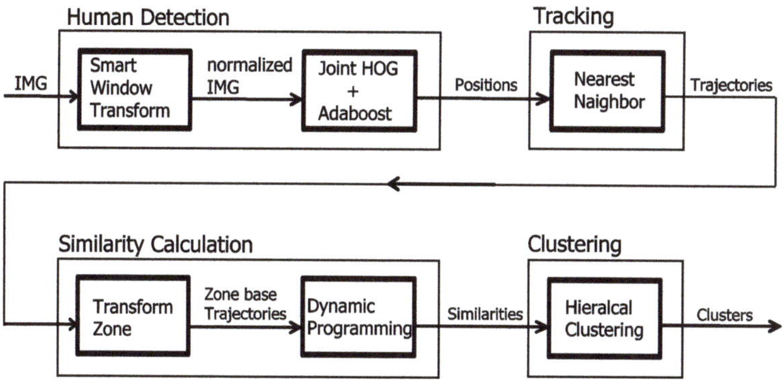

Fig. 1.1 Procedure of clustering method for zone based trajectories

Fig. 1.2 Detection window using SWT and normalized image

for local features selected from thousands training samples, such as Edge Orientation Histogram (EOH) [2] and Histograms of Oriented Gradients (HOG) [3], combined with AdaBoost, SVM or so on.

After the training process, human detection is carried out by raster scanning a observed image with all rectangular detection windows in the image coordinate.

When cameras are either distant or perspective distortion is not obvious, it works well. However, in the real environment, camera setting (camera position, angles, height, etc.) is random and difficult to control.

To solve these problems, we propose a new method, called Smart Window Transform (SWT). SWT can construct a robust detection window to camera setting and obtain more suitable human silhouette or counters for classification.

Fig. 1.3 Zone identification

In this paper, we use Joint-HoG and Adaboost [4]. A novel method called Joint-HoG is per- formed with 2-stage Adaboost. This method can automatically capture human shape symmetry and continuity, and combine HOG features for several different local areas by means of the first-stage boosting. Then the generated joint features are input to the second-stage boosting, to construct the final classifier and detect the objects. Thanks to this method, the continuous shape from head to shoulder can be captured efficiently. Also, the symmetry of human head, shoulders and legs, etc. can be selected.

At first, for any observed position in the world coordinate, SWT is performed to get a detection window with the best shot of human and transfer it into the image coordinate. Then normalization is performed to transfer this detection window into a rectangular to fit a human classifier. Later, a edge-based classifier is used to extract the local features inside normalized window and judge whether it is human or not. At last, all detection results are presented and clustered in the world coordinate using MeanShift [5].Trajectory is defined as time sequential position detected in the world coordinate (Fig. 1.3).

1.2.2 Definition of Zone

Shopper's purpose of action is to buy merchandises, so we should consider interaction of shopper and merchandises for more effective analysis. The areas where shopper can interact with merchandises are defined as "Zone" and trajectories are analyzed by it. In this section, we describe the definition of Zone and the

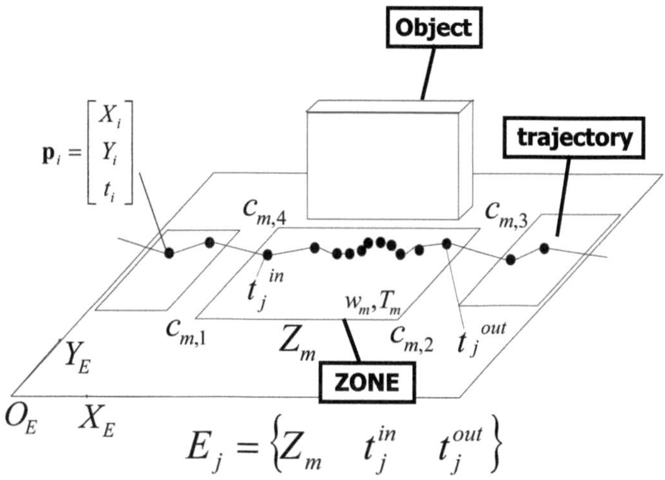

Fig. 1.4 Definition of Zone

expression of trajectory based on Zone using Fig. 1.4. A position of a shopper $\mathbf{p_i}$ and trajectory P is defined by the expression below. Zone Z is defined in the real world space.

$$\mathbf{p}_i = [X_i \; Y_i \; t_i]^T \tag{1.1}$$

$$P = \{\mathbf{p}_i\}_{i=1,2,\ldots,I} \tag{1.2}$$

$$Z = \{\mathbf{c}_1, \mathbf{c}_2, \mathbf{c}_3, \mathbf{c}_4, w, T\} \tag{1.3}$$

$$\mathbf{c}_n = [X_n \; Y_n]^T \tag{1.4}$$

The parameter $\mathbf{c_n}$ is each corner point of Zone, X_n and Y_n is real world coordinates, w is weight and T is a threshold for discriminating the stop. Z_m is defined as the Zone which shopper trajectories and the time of enter/exit as t^{in}, t^{out} of a ENTRY E. ENTRY base trajectory P^E is expressed by time sequential ENTRY and defined by expression, below equations.

$$E = \{Z_m, t^{in}, t^{out} \mid 1 \le m \le M\} \tag{1.5}$$

$$P^E = \{E_j\}_{j=1,2,\ldots,J} \tag{1.6}$$

Shopper stops in the Zone shows interest merchandises. So the shoppers who stop the same Zone have the same interest for the merchandises. From this, information of the stop of a shopper is very important for analysis of shoppers' behavior. The parameter t^{in} is the time of entering a Zone and t^{out} is the time of exit. The Entry that the time of stop time $t^{out} - t^{in}$ is longer than T is defined as Stop Entry. The

Fig. 1.5 Shopper trajectory detected manual

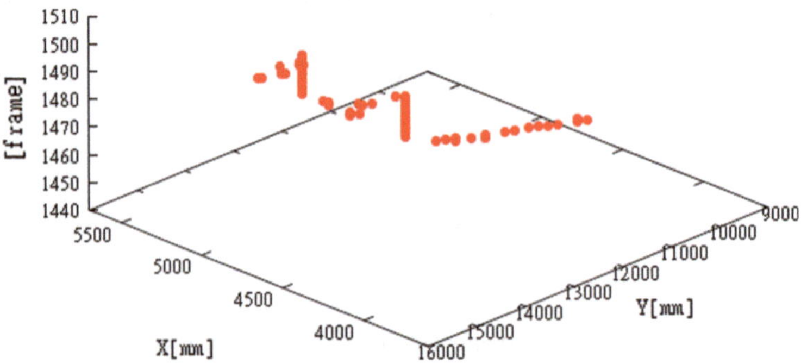

Fig. 1.6 Spatiotemporal trajectory expression of shopper's trajectory

trajectory composed by Stop Entries is Stop Entry base Trajectory P^{SE} and it is defined by below equations (Figs. 1.5, 1.6, 1.7 and 1.8).

$$E^S = E \mid t^{out} - t^{in} \geq T_m \tag{1.7}$$

$$P^{SE} = \{E_1, \{E_k^S\}_{k=1,2,\ldots,K}, E_J\} \tag{1.8}$$

1.2.3 Similarity Calculation Method of Zone Base Trajectories

We define a new method called "similarity calculation method of Zone base trajectories" to calculate the similarity of ENTRY base trajectories P^{SE}.

Fig. 1.7 Shopper trajectory detected manual

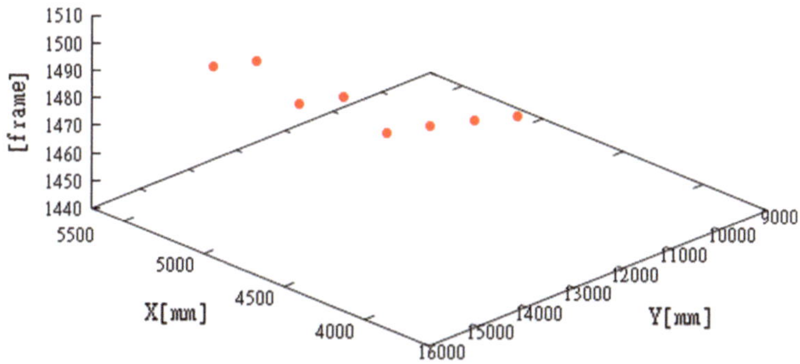

Fig. 1.8 Spatiotemporal trajectory expression of shopper's trajectory

Length of ENTRY base trajectories is different to each other, so positions which trajectories is incomparable to each other. To solve this problem, we use Dynamic Programming (after DP [6]) for matching of time sequential Zone data in this paper. DP is used in voice recognition. This method can search optimum matching of data that have different length. Similarity calculation method of Zone base trajectories is defined as below. Parameter $r(i,j)$ is weight for insertion of data. d_{tra} is distance, d_{ins} is cost of insert, α, β is adjustment factor, $d(i,j)$ is distance of i and j plus optimum data of one step before, $s_{p,q}$ is similarity of trajectory P and Q, $d_{p,q}$ is dissimilarity and d_{max} is maximum of dissimilarity.

$$d_{tra}(i,j) = \left|E_i(Z_n(c_1)) - E_j(Z_m(c_1))\right|^{\alpha} \tag{1.9}$$

$$d_{ins}(i,j) = r(i,j) \cdot \beta \tag{1.10}$$

$$d(i,j) = d_{tra}(i,j) + \min \begin{cases} d(i-1,j) + d_{ins}(i,j) \\ d(i,j-1) + d_{ins}(i,j) \\ d(i-1,j-1) \end{cases} \tag{1.11}$$

$$d_{p,q} = d(I,J) \tag{1.12}$$

$$s_{p,q} = \frac{d_{max} - d_{p,q}}{d_{max}} \cdot 100 \tag{1.13}$$

1.2.4 Integration of Image and ID-POS

Calculate similarity of trajectories made from ID-POS and made from image. The method that detect trajectory from ID-POS is defined as POS2ZT. POS consist of sales merchandises m_i and purchase time t.

$$POS = (\{m_i\}, t) \tag{1.14}$$

Function that transform merchandise m to store shelf G and transform store shelf G to zone Z is defined as below.

$$f_{GM}(m) = G \tag{1.15}$$

$$f_{ZG}(G) = Z \tag{1.16}$$

Then, function that transform merchandise to zone is defined as below.

$$\begin{aligned} f_{ZM}(m) &= f_{ZG} \cdot f_{GM}(m) \\ &= Z \end{aligned} \tag{1.17}$$

Transformed POS using $f_{ZM}(m)$ don't have order. Make POS Zone-based Trajectory (POSZT) by connecting all combination of sales merchandises. By calculating similarity of POSZT and Zone-based trajectory, they are identified each other.

1.3 Experiments of Identify Shopper and ID-POS

The experiment is cooperated with COOP Sapporo. COOP Sapporo has 107 shops in Hokkaido. The experimental shop is a drag store in the COOP Sapporo of Yoichi. Making a detection window using camera image parameter of camera is necessary.

1.3.1 Behavior Analysis of ISZOT

Trajectories are detected from images that are taken in COOP Sapporo of Yoichi at 10:00 to 12:00 on September 15th, 2012. The number of trajectories is 195

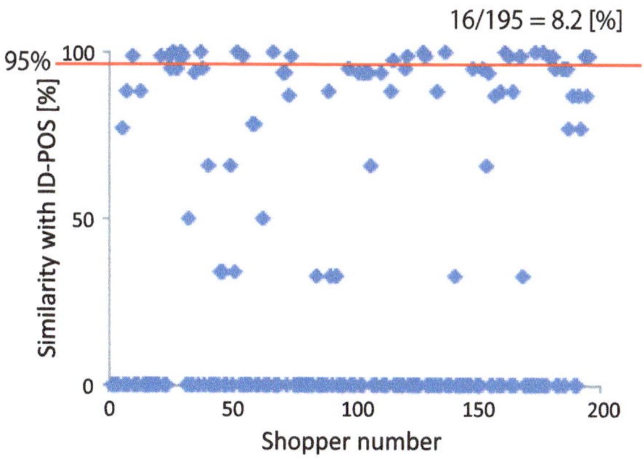

Fig. 1.9 Simirarity of POS which buy merchandise at Zone 4 and each shopper who have been at the area

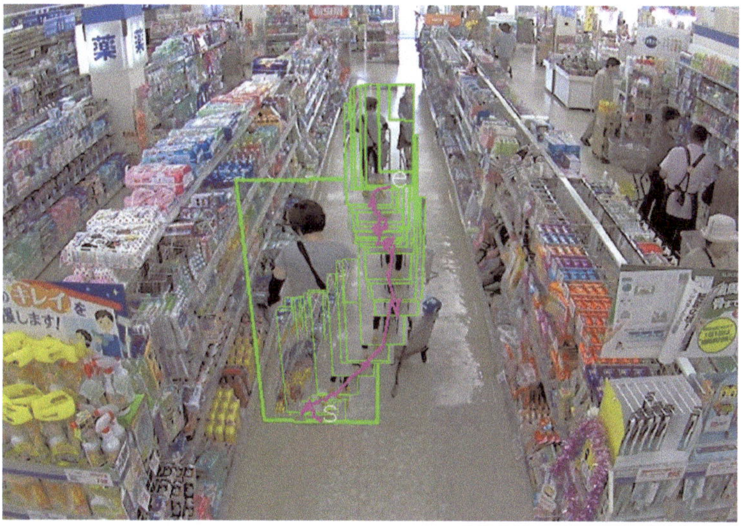

Fig. 1.10 Shopper's trajectory which is identified with a POS data

shoppers. The number of ID-POS, which is taken by COOP Sapporo, is 27 shoppers on the same day and time with trajectories. Figure 1.9 shows similarity of a ID-POS which recorded the merchandises in Zone 4 and trajectories. In this result, the number of shoppers who has similarity higher than 95 % is 16 shoppers. Our method can reduce 91.8 % of searching range. Then shopper's detention can be divided into buying and only considering, due to identify ID-POS with trajectory. Buying is behavior that buys after detention, and only considering is behavior that not buys after detention (Figs. 1.10 and 1.11).

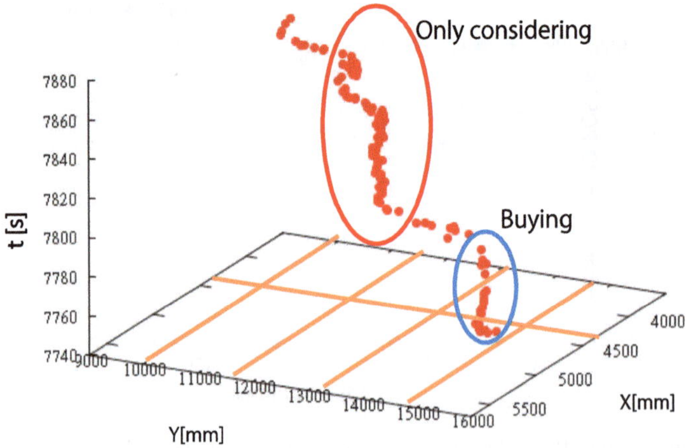

Fig. 1.11 Shopper's time and space trajectory view which is identified with a POS data

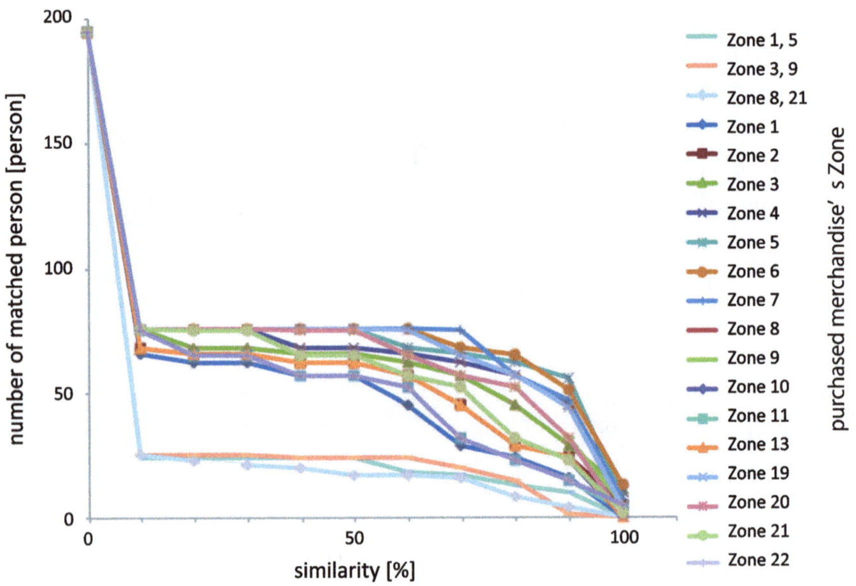

Fig. 1.12 Narrow down a searching range

1.3.2 Narrow Down a Searching Range

Figure 1.12 shows the number of shoppers who have higher similarity than threshold about the 27 ID-POS data calculated similarity in sequence 1.13. From result of experiment, we understood follows.

- When threshold of similarity is 90 %, our method can reduce searching range to average 87.6 %.
- When threshold of similarity is 10 %, our method can also reduce searching range to average 65.9 %.
- ID-POS, which records more than two zones, converges earlier than which records only one zone.
- It may not be become one-to-one.
- It can't be identified if shoppers do same behavior, because our method analyze shoppers from trajectories.

1.4 Conclusions

In this paper, we explained the system for the shopper behavior analysis system-using image based sensing named ISZOT. Clustering method of Zone base trajectories was proposed. This paper proposed three main methods, human detection method using SWT, similarity calculation method of Zone base trajectories and integration of image and ID-POS. We get video of security camera from COOP Sapporo of Yoichi for two months. Using human detection and tracking method, we got 195 shoppers' trajectories. In addition we showed experiment to detect ID-POS data from 195 shoppers' trajectory. Our method can reduce searching range to average 87.6 % when threshold of similarity is 90 %. Even if the threshold of similarity is 10 %, our method can also reduce searching range to average 65.9 %.

By reducing searching range, it becomes easily to associate trajectory with ID-POS one by one by human eye. If we can do it, shopper's detention can be divided into buying and only considering. It let us know deep understanding for shoppers' behavior, and the effective way of advertisement or efficient product arrangement.

Acknowledgments We would like to thank Keitaro YONEDA from CO-OP Sapporo and Takayuki TANAKA from Hokkaido University for their cooperation of experiment data and valuable discussion.

References

1. Mohammed N, Fung B, Debbabi M (2009) Walking in the crowd: anonymizing trajectory data for pattern analysis. In: Proceedings of the 18th ACM conference on information and knowledge management, pp 1441–1444
2. Levi K, Weiss Y (2004) Learning object detection from a small number of examples: the importance of good features. In: Proceedings of the 2004 IEEE Computer Society conference on computer vision and pattern recognition. CVPR 2004, vol 2, p II–53
3. Dalal N, Triggs B (2005) Histograms of oriented gradients for human detection. In: IEEE Computer Society conference on computer vision and pattern recognition. CVPR 2005, vol 1, pp 886–893

4. Mitsui N, Fujiyoshi H (2009) Object detection by joint features based on two-stage boosting. In: IEEE 12th international conference on computer vision workshops (ICCV Workshops), pp 1169–1176
5. Cheng Y (1995) Mean shift, mode seeking, and clustering. IEEE Trans Pattern Anal Mach Intell 17(8):790–799
6. Bertsekas D (1995) Dynamic programming and optimal control, vol 1. Athena Scientific, Belmont

Chapter 2
Low-Cost IR Visualizer Based on a Rotating Phosphor Screen for Accurate Beam Analysis

Jaebin Chang, Jangbeom Lee and Sucbei Moon

Abstract We present a low-cost IR beam analysis tool based on a rotating phosphor screen. The infrared light incident on the IR-sensitive screen was converted to visible light by the IR-responsive phosphor that operated by light charging and IR-induced luminescence. Through comparative experiments, the conventional IR viewing card, based on the same kind of phosphor, was found to exhibit a short luminescence time because the excited fluorescent molecules are depleted locally at the beam spot. This effect hinders accurate estimation on the spatial characteristic of the IR beam and consequently limits the capability of the IR viewing card in beam inspection. Our IR visualizer alleviates this effect by rotating the IR-sensitive screen in a high speed so that the luminescence time was effectively prolonged to allow a longer measurement time for the IR beam. Our scheme provides a simple and economic means of IR beam inspections, especially useful for the wavelength band above 1,000 nm where the solid-state image sensors are hardly available in low cost.

Keywords Infrared analysis • IR viewing instrument • Beam analysis • IR-sensitive phosphor

2.1 Introduction

Infrared (IR) light is invisible electromagnetic radiation with a wavelength longer than that of visible light. Near-IR radiation of relatively short wavelengths, ranging from 700 nm to 1,700 nm, is particularly of practical interest in applications such as

J. Chang
Hankuk Academy of Foreign Studies, Yongin-si, South Korea

J. Lee · S. Moon (✉)
Kookmin University, Seoul, South Korea
e-mail: moons@kookmin.ac.kr

R. Tutsch et al. (eds.), *Progress in Optomechatronic Technologies*,
Lecture Notes in Electrical Engineering 306, DOI: 10.1007/978-3-319-05711-8_2,

optical communications, biomedical imaging and optical measurements due to the wide availability of light sources and transparent materials in the near-IR band. It is widely demanded to visualize the near-IR beams in those application fields for testing the basic properties of the IR beam [1–5]. An IR viewer that converts the IR radiation to a visible image of the beam can be used for laser safety or beam characterizations as a routine tool of development and maintenance [2–5]. An IR viewing card is the most popular means for detecting the invisible IR beam in an economic way. The phosphorous screen of the IR viewing card emits visible fluorescence in response to the incident IR radiation and easily visualizes the spatial pattern of the IR beam [2, 3]. Compared to a heavier instrument of an IR viewing scopes which combines an IR camera and a display unit in an instrument [5], the IR viewing card is much cheaper and easier to use for practical IR beam inspections. It has been widely utilized in the near-IR applications when the presence of the IR light needs to be checked with rather qualitative information on the spatial distribution of the light fields. Still, accurate and quantitative beam analysis requires IR-sensitive cameras such as a more expensive InGaAs image sensor for the wavelengths above 1,000 nm.

For IR-responsive luminescence, the screen of the IR viewing card must rely on a multi-photon process. A single IR photon alone cannot produce visible light because of the lower energy of the IR photon. The principle of the IR sensing is far different from ultraviolet-to-visible conversion which is easily achieved by the fluorescence of a partially energy-dissipating process. Two different types multi-photon processes can be used for the purpose of the IR-to-visible conversion in the IR viewing card: up-conversion and light charging. In up-converting phosphor, more than two low-energy photons are absorbed virtually simultaneously by help of the intermediate states [6]. The final high energy is released at once to make a visible photon by fluorescence. The up-conversion IR viewing card is convenient to use but has several disadvantages. It usually has a narrow spectral range of IR detection with a relatively low conversion efficiency. The IR sensitivity is frequently unacceptably low for typical up-converting phosphors when the intensity of the IR is relatively low. The non-proportional response of the nonlinear process also degrades the applicability of this method, which distorts the spatial characteristic of the detected IR beam significantly.

In the IR viewing card based on light-charging phosphor, a visible or ultraviolet photon is absorbed at first to excite the phosphor molecules in a meta-stable state of a very long lifetime [7]. An IR photon that comes after the energy charging triggers the energy release with a visible fluorescence which signifies the IR light. The light-charging IR viewing card, thus, requires a procedure of light charging before being used for the IR detection. Due to the wide wavelength range of a single card and a high sensitivity of the IR detection capability, this type of IR viewing cards are popular in the near-IR applications in spite of the notable inconvenience of light charging requirement. The real drawback of the light charging IR viewing card is found in the nature of *discharging effect*. Because a limited number of molecules are involved in the IR sensing, the excited electrons become depleted by the luminescence process. The card only maintains the luminescence brightness for a short time, typically for a few seconds when exposed to an intense IR beam. Once discharged, its conversion efficiency does not recover unless it is recharged with intense visible light. Shaking the IR viewing card by the user's hand can help extend the luminescence time but hinders quantitative estimation of the beam characteristic in actual use. After all, the IR viewing card can just provide a limited chance to check the approximate position and the size of the IR beam. A better IR visualizing means is highly demanded for various inspection purposes with capabilities of rather quantitative IR beam analysis.

In this report, we present a new scheme of IR visualization that greatly improves the performance of the light-charging IR viewing card. In our scheme of the rotating IR-sensitive screen, the IR-responsive screen based on the light-charging phosphor is rotated in a high speed to increase the exposure area of the IR beam and, hence, to increase the luminescence time with a reduced local depletion of excited molecules. Despite the simple principle, it effectively allows measurements of the IR beam in a longer time with reduced distortion of the detected spatial pattern. More quantitative analysis of the IR beam is enabled in such a very simple and economic method. We have experimentally tested the feasibility of the IR visualizer based on the rotating IR-sensitive screen and have found that our scheme can provide an inexpensive but effective scheme of IR beam analysis.

2.2 Experiment

In this research, a commercially available IR viewing card that requires visible light charging for IR detection (VRC5, Thorlabs Inc.) was utilized [2]. Figure 2.1 shows the approximate IR detection sensitivity obtained from the manufacturer's datasheet. The intensity of the visible luminescence is plotted as a function of incident IR wavelength. The IR detection range was so wide that a single IR viewing card could cover the whole near-IR band. Once charged by visible light, the phosphor screen could detect the IR light for a long latency time, longer than tens of minutes, and emitted reddish orange light in response to incident IR radiation.

Our IR visualizer was made of an IR-sensitive screen mounted on an electric motor that can rotate the screen stably. The IR detection wheel of a circular plate was prepared so that half of the wheel was covered with the IR-sensitive screen obtained from the commercial IR viewing card (VRC5) while the other half was covered with a non-fluorescent orange plate. The diameter of the wheel was 36 mm in full. A super-luminescent LED was placed above the wheel and could be used for light charging with the intense white light. The screen was uniformly charged before being used by turning the LED on while the wheel was rotating. Figure 2.2 shows our IR visualizer in an experimental setup for testing the feasibility. A collimated IR beam from a fiber-optic light source was launched to the IR detection wheel with an angle of 61° with respect to the plane of the wheel. The wavelength and the beam diameter of the IR beam were 1,310 nm and 2.09 mm, respectively. A digital camera built with a silicon image sensor was used to capture the IR-induced luminescence pattern, which represents the shape of the IR beam projected onto the IR-sensitive screen.

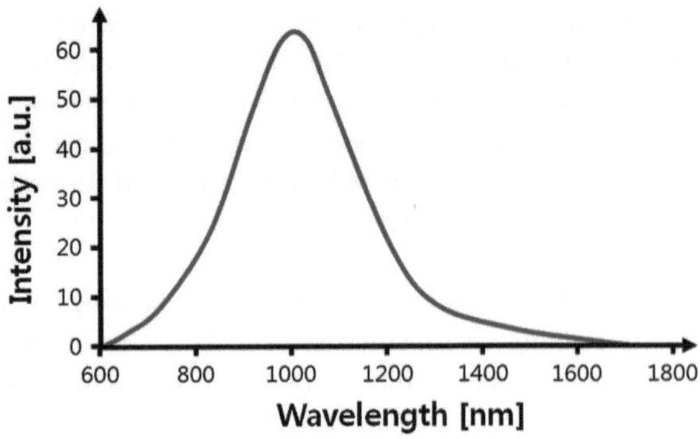

Fig. 2.1 IR-induced luminescence intensity versus wavelength of the incident IR for the IR-sensitive screen used in our experiment [2]

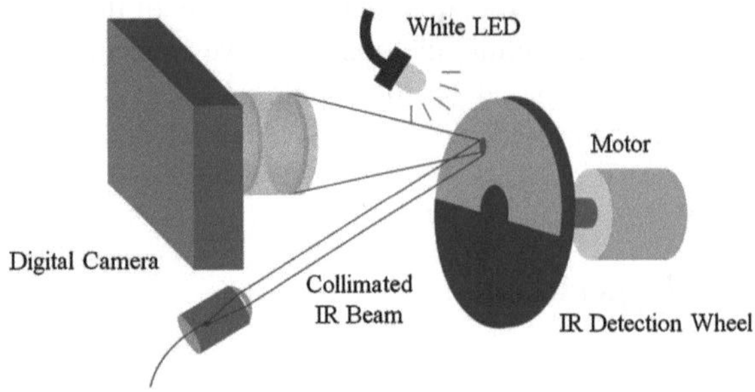

Fig. 2.2 Experimental setup for testing the IR visualizer

After charging the screen, the conversion of IR to visible light was better observed with a higher contrast by turning off the charging light. The principle of our IR visualizer is to increase the effective detection area by rotating the wheel of the IR-sensitive screen. For a point of the screen, the IR exposure time is greatly reduced by such a scanning mechanism. Filling the half of the wheel only with the active phosphorus screen, the visible pattern automatically flickered in a high frequency due to the rotation. It could help the user sense the IR response with better awareness. Flickering luminescence was easily distinguished from ambient light or a visible pattern occasionally incident upon the screen.

A comparative experiment was performed to evaluate the improvement in IR detection made by our scheme compared to an IR viewing card in a fixed position. This was performed for two different cases with the same setup. In the case of the *rotating IR detection wheel*, the electric motor rotated the wheel at a speed of ~1,000 r.p.m. In the case of the *fixed IR detection wheel*, the wheel was fixed so that the IR beam pointed at a constant position on the active IR-sensitive screen on the wheel. The latter case corresponded to the conventional use of the IR viewing card. For each case, three different intensities of IR rays, 0.55, 1.81 and 5.07 mW for an effective beam area of 0.27 cm^2, were applied to the IR detection area and were recorded for comparison. Then, the intensity and the beam width of the IR-induced visible pattern were evaluated from captured images of the IR-induced luminescence.

The experiment was carried out in the following steps. For the fixed IR detection wheel, the IR-sensitive screen was charged fully by being exposed to the white light of the LED. The screen was left in dark environment for 30 s after the charging was finished. Next, the IR beam was launched onto the screen while the camera recorded the IR-induced visible pattern that the screen produced. These procedures of charging and IR irradiation were repeated with different IR powers in the same condition and procedures. As well, this experiment was also performed for the rotating IR detection wheel case in the same way. To obtain a wider dynamic range of the measurement, the exposure time of the camera was adjusted in keeping the frame rate at 5 frames/s so that no pixel was saturated. The recorded digital images were processed in a computer to find the intensity of the IR-induced luminescence at the center of the beam. The measured pixel value of luminescence was normalized by the camera's exposure time for each measurement. The change of the relative intensity in time was compared for the different cases. The apparent beam width was also evaluated in full width at half maximum (FWHM) along an axis from the acquired images. And its change in time was also analyzed to find beam measurement errors of the spatial characteristic.

2.3 Result and Discussion

We have analyzed the spatial distributions of the IR-induced visible patterns through the images taken by the digital camera. Its variation in time, t, was evaluated to find the effect of discharging for the two cases. In the description of the experiment results, $t = 0$ was set to be the time of starting the IR irradiation. Figure 2.3 shows the images and contour map representation of the luminescence intensity for the two cases: (a), (b) and (c) for the case of the fixed wheel; (d), (e) and (f) for the case of the rotating wheel. The plots of Fig. 2.3g, h show the intensity distributions along the vertical dotted lines of the contour maps for those two cases, respectively. In detail, Fig. 2.3 shows the image of the acquired visible pattern at $t = 2$ s (a), its contour map representation (b) along with the contour map of the visible intensity at $t = 8$ s (c), respectively. For the rotating case, it also shows the image at $t = 2$ s (d), the intensity-based contour map at $t = 2$ s (e) and $t = 20$ s (f). The inset square in Fig. 2.3a or d gives a scale factor of 1 mm $\times$ 1 mm. Note that the circular IR beam was projected onto the screen with an incidence angle of 29° which resulted in the oval shapes of the visible patterns. As a consequence, the circular cross-section of the beam was slightly extended along the horizontal axis in those images.

The images and contour maps of Fig. 2.3 suggested that the fixed wheel case suffered from the drastic decrease in intensity of the IR-induced luminescence in a few seconds after the IR irradiation started. In addition, the spatial distribution of the IR intensity was incorrectly represented when the discharging effect considerably reduced the visible luminescence. It is easily explained by the fact that the center of the beam experienced the fastest depletion of the excited molecules due to the highest IR intensity. This effect obviously made it difficult to find the center

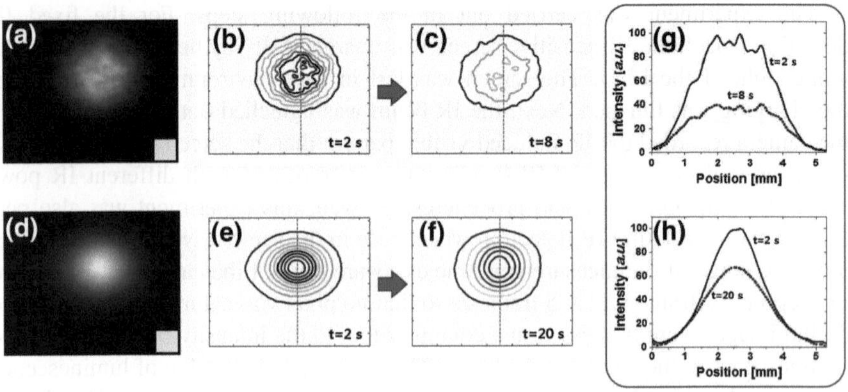

Fig. 2.3 Visible pattern of the IR beam at t = 2 s (**a**), its contour map representation (**b**), that of the visible intensity at t = 8 s (**c**), respectively, for the case of the fixed IR detection wheel; shown with the visible pattern acquired at t = 2 s (**d**), the intensity contour maps acquired at t = 2 s (**e**) and t = 20 s (**f**) for the case of the rotating IR detection wheel, respectively. The intensity distributions along the *vertical dashed lines* of the contour maps are plotted in (**g**) and (**h**)

of the beam and the beam width with the IR viewing card. In contrast, the rotating wheel provided much better representation of the IR beam with reduced discharging effect as clearly observed in Fig. 2.3e, f. One more advantage of the rotating wheel is found in its smooth images. The images of the fixed wheel directly reflected the roughness of the screen surface as seen in Fig. 2.3a. The rotating wheel smoothened the irregularity of the surface by the fast rotation and gave clearer representation of the IR intensity distribution as observed in Fig. 2.3d. The Gaussian-like beam characteristic was successfully measured with our IR visualizer as observed in Fig. 2.3.

The decrease of the IR-induced luminescence at the center of the beam in time was evaluated from the image data. For better precision, the luminescence values of the neighboring pixels in the vicinity of the center were averaged to find the *central intensity*. Figure 2.4 shows the central intensity of the IR-induced luminescence as a function of time, t, for the case of the fixed IR detection wheel (a) and that of the rotating wheel (b), respectively. For the fixed case, discharging effects were clearly observed with the nearly exponentially decaying curves. The luminescence time was defined as the time of luminescence by which the IR-induced luminescence decays to the half of the peak intensity. It was measured to be 2.7 s with an incident IR power of 5.07 mW, 6.1 s with a power of 1.81 mW, and 19.6 s with a power of 0.55 mW. The rotating IR detection wheel also exhibited a decaying characteristic but with a much longer luminescence time. It was measured to be 35 s for the incident IR power of 5.07 mW. It was longer than 80 s when the IR power was less than 1.81 mW. This observation demonstrated our scheme extends the effective measurement time by a factor more than ten compared to the conventional IR viewing card.

Discharging or the decay of the IR-sensitive luminescence does not only decrease the luminescence intensity but also disturbs the accurate measurement of the beam shape and the position determination. The difference in decay speed

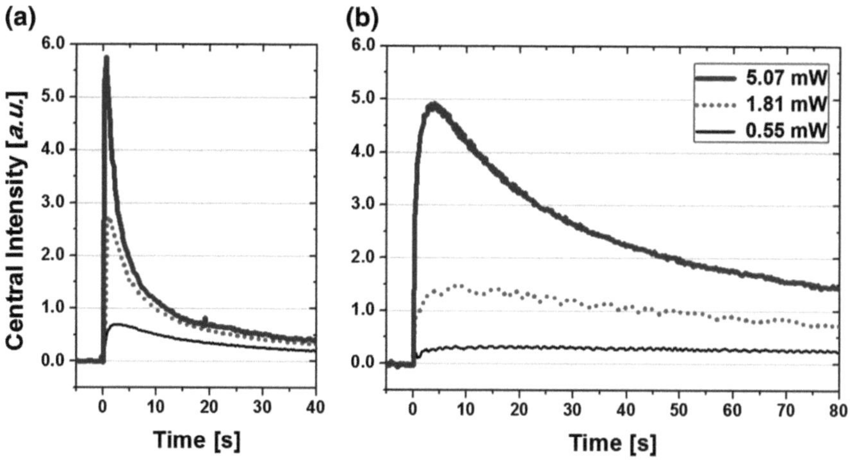

Fig. 2.4 Central intensity of the IR-induced luminescence as a function of time, *t*, for the case of the fixed IR detection wheel (**a**) and that of the rotating wheel (**b**), respectively

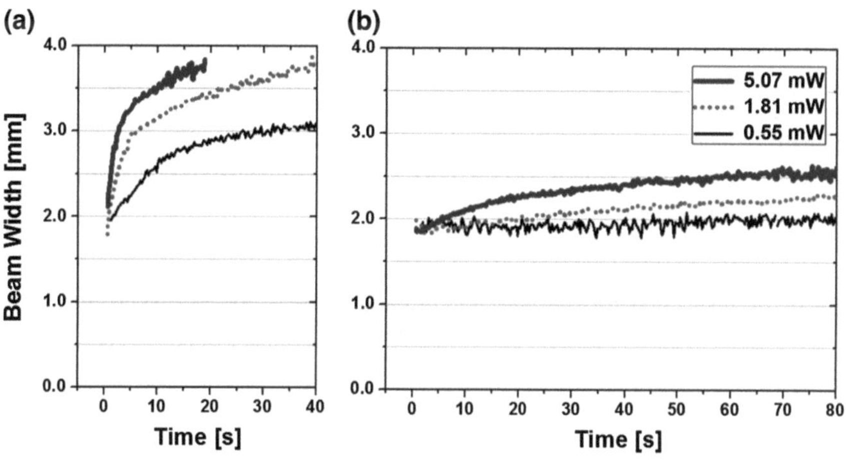

Fig. 2.5 Measured beam width of the IR-induced visible pattern of the fixed wheel case (**a**) and that of the rotating wheel case (**b**), evaluated in FWHM, respectively

is explained by the different IR intensity for each position of the screen. As the central region decreases fastest in luminescence intensity, the beam width extends accordingly. From the image data, the beam width was measured long the vertical axis of each image. Figure 2.5 shows the beam width of the IR-induced visible pattern of the fixed wheel case (a) and that of the rotating wheel case (b), measured in FWHM, respectively. Note that the actual beam width of the IR beam was measured to be 2.09 mm. For the case of the fixed IR-sensitive screen, the measured beam width increased very rapidly as the discharge of the phosphor screen. Even for a low IR power less than 1 mW, it could not be in an acceptable range of error unless the measurement was done in a few seconds after the IR irradiation started. In contrast, the rotating IR-sensitive screen gives much more accurate results within a measurement time of half a minute. It suggested that accuracy and precision could be enhanced by our scheme of the rotating screen with a better representation of the spatial characteristic of the IR beam.

2.4 Conclusion

In this report, we present a low-cost IR beam analysis tool based on a rotating IR-sensitive phosphor screen. The infrared light incident on the IR-sensitive screen was converted to visible light by the phosphor that operated by light charging and IR-induced luminescence. The measurement range of IR wavelength was so wide, from 600 through 1,600 nm, that the entire band of the near-IR region was covered at once. Through comparative experiments were performed to compare our scheme with the conventional IR viewing card based on the same kind of phosphor. It was found that the stationary use of the IR-sensitive screen exhibits a very short luminescence time because the excited fluorescent molecules are depleted locally at the beam spot. This effect was found to hinder accurate estimation on the spatial characteristic of the IR beam. It limits the capability of the IR viewing card in beam inspection. Our IR visualizer alleviates this effect by the rotating IR-sensitive screen in a high speed so that the luminescence time was effectively prolonged to allow a longer measurement time for the IR beam. The experimental result demonstrated that our scheme provides a simple and economic means of IR beam inspections. This new technique of IR beam characterization can be a useful tool of IR inspection for the wavelength band above 1,000 nm where the solid-state image sensors are hardly available in low cost.

References

1. Derickson D (1998) Fiber optic test and measurement. Prentice Hall PTR, New Jersey
2. Laser Viewing Cards, Thorlabs Inc. http://www.thorlabs.com/newgrouppage9.cfm?objectgroup_id=296&pn=VRC5#7125
3. Fonda GR, Schenectady NY (1948) Infrared-responsive phosphors. United States Patent No: 2, 447–322
4. Ebitani M, Tominaga T, Kishi A (1995) Infrared-to-visible converter. United States Patent No: 5, 438–198
5. IR and UV Viewing Scopes, Edmund Optics Inc. http://www.edmundoptics.com/lasers/laser-measurement/infrared-ir-ultraviolet-uv-viewers/ir-uv-viewing-scopes/1402
6. Auzel F (2004) Upconversion and anti-stokes processes with f and d ions in solids. Chem Rev 104:139–173
7. Goldstein B, Dropkin JJ (1962) Infrared sensitivity of the ZnS:Cu:Co phosphor. Phys Rev 126:966–970

Chapter 3
Evaluation of Heteroepitaxially Grown Semipolar {20-21} GaN on Patterned Sapphire Substrate

Yasuhiro Hashimoto, Masakazu Koyama, Takashi Inagaki, Keisuke Yamane, Narihito Okada and Kazuyuki Tadatomo

Abstract Semipolar {20-21} GaN layers were grown on {22-43} patterned sapphire substrates by metal–organic vapor phase epitaxy using a two-step growth method. We succeeded in suppressing −*c*-plane growth at growth temperatures of 1,000 and 900 °C for the first and second steps, respectively; the resulting structure exhibited a large reduction in the number of stacking faults upon optimizing the growth conditions. Photoluminescence measurements showed an increase in the near-band-edge emission and a decrease in deep-center emission when the two-step growth was performed at higher V/III ratio.

Keywords MOVPE • GaN • PSS • {20-21} • Stacking fault • PL

3.1 Introduction

GaN and other related semiconductor materials have been widely used in the fabrication of light-emitting diodes (LEDs), laser diodes (LDs), radio-frequency devices, and power devices. In general, commercially available LEDs and LDs are fabricated on a polar *c*-plane GaN. However, the quantum confined Stark effect (QCSE) due to polarization-induced strong internal electric fields inside *c*-plane InGaN/GaN multiple quantum wells causes the degradation of device performance. QCSE is reduced by using semipolar and nonpolar GaN substrates [1]. Most notably, the green LD has been realized by using a semipolar {20-21} GaN substrate [2], and improvements in the efficiency droop of LEDs for high injection currents have been achieved by using a semipolar {20-2-1} GaN substrate [3]. A nonpolar or semipolar GaN substrate is usually fabricated by slicing

Y. Hashimoto · M. Koyama · T. Inagaki · K. Yamane · N. Okada · K. Tadatomo (✉)
Graduate School of Science and Engineering, Yamaguchi University, Yamaguchi, Japan
e-mail: tadatomo@yamaguchi-u.ac.jp

R. Tutsch et al. (eds.), *Progress in Optomechatronic Technologies*,
Lecture Notes in Electrical Engineering 306, DOI: 10.1007/978-3-319-05711-8_3,

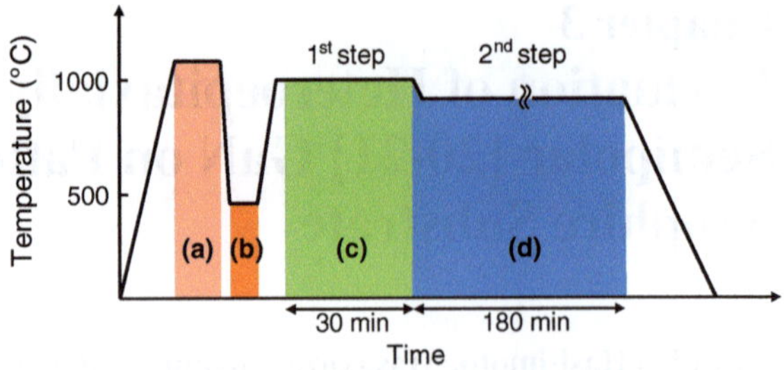

Fig. 3.1 Schematic of temperature profile for (**a**) thermal cleaning (**b**) growth of low-temperature GaN buffer layer **c** growth of GaN layer in first step, and (**d**) growth of GaN layer in second step. The V/III ratio was varied from 147 to 2,942

bulk *c*-plane GaN. Hence, large-sized semipolar {20-21} GaN substrates are not commercially available to grow the abovementioned device structures. Therefore, the fabrication of large-sized {20-21} GaN substrates on foreign substrates as well as *a*- [4], *m*- [5, 6], {10-1-3}, {11-22} [7], and {10-11} [8] is desirable. However, the use of {20-21} GaN templates on foreign substrates such as sapphire or silicon has not thus far been reported. We have previously reported the fabrication of a 2-inch-diameter {20-21} GaN layer on a patterned sapphire substrate (PSS) by using metal–organic vapor phase epitaxy (MOVPE) [9] and the subsequent fabrication of a 2-inch-diameter freestanding {20-21} GaN by using hydride vapor phase epitaxy (HVPE) [10]. In this light, previous studies have reported on the heteroepitaxial growth mode of semipolar GaN [11–13]. In order to obtain a high-crystalline-quality bulk {20-21} GaN substrate using the sapphire substrate, the crystalline quality of the {20-21} GaN layer is also important, and its growth mode should be investigated in detail. However, the growth mode of the {20-21} GaN layer has not thus far been investigated. In this work, we investigated the heteroepitaxial growth mode of a {20-21} GaN layer on PSS and the resulting crystalline quality of the {20-21} GaN layer.

3.2 Experimental Procedure

For the growth of the semipolar {20-21} GaN layer, we prepared a stripe-patterned {22-43} PSS with an SiO_2 mask. Stripe patterns were formed perpendicular to the *c*-axis of {22-43} sapphire by using conventional photolithography and inductively coupled plasma reactive ion etching (ICP-RIE). As a consequence, the stripe-patterned {22-43} PSS had a *c*-plane-like sapphire sidewall, because it is difficult to fabricate the exact *c*-plane sapphire sidewall using ICP-RIE. The depth and width of the grooves formed by etching were 1 and 3 μm, respectively. The period of the stripe pattern was 6 μm. The SiO_2 mask was deposited on the ridges to prevent the growth of GaN [13]. A {20-21} GaN layer was grown on the PSS by using MOVPE. Trimethylgallium (TMG) and ammonia (NH_3) were used as the sources of Ga and N, respectively. The carrier gas used was purified H_2. Figure 3.1 shows the temperature profile corresponding to the growth of a {20-21} GaN layer by MOVPE. A low-temperature GaN buffer layer was grown at 460 °C after thermal cleaning. Subsequently, a {20-21} GaN layer was grown by using a two-step growth method. The two-step growth method is useful to obtain perfectly coalesced GaN layers on the {22-43} PSS. Crystal growth was not interrupted between the first and second steps.

The {20-21} GaN layers were observed by scanning electron microscope (SEM) with an acceleration voltage of 5 kV. Stacking faults were characterized by photoluminescence (PL) measurements at 4 K by using a 14 mW, 325 nm He–Cd laser.

3.3 Results and Discussion

Firstly, we investigated the one-step growth of the semipolar {20-21} GaN layer on the {22-43} PSS [9]. The growth temperature and the V/III ratio were 1,000 °C and 294, respectively. This growth condition was optimized to obtain a selective area growth from the *c*-plane-like sapphire sidewall. Figure 3.2a, b show the cross-sectional SEM images of GaN layers grown for 60 and 300 min, respectively. The initial facet structure was composed of one $-c$-plane and two {10-11}, as shown in Fig. 3.2a. However, the GaN layers did not coalesce with each other after 300 min of growth. Furthermore, the $-c$-plane GaN was grown as shown in Fig. 3.2b, with the growth region exhibiting a large defect density [14, 15]. These defects were mainly characterized as the I_1 type of basal stacking faults (BSFs) by TEM observations. To mitigate these problems, we adopted the two-step growth method in this study.

The growth of the GaN layer at lower temperatures was effective in enhancing the growth rate toward the $+c$-direction such that the $+c$ facet structure shown in Fig. 3.2b disappeared [16]. Therefore, the two-step growth method was adopted to obtain coalesced continuous GaN layers. The growth temperatures of the first and second steps were 1,000 and 900 °C, respectively. The V/III ratio was maintained at 294 during the two-step growth process. The first step initiates the growth of the GaN nucleus (Fig. 3.2a), and the second step leads to coalescence of the GaN layers. Figure 3.3 shows a cross-sectional SEM image of a perfectly coalesced GaN layer achieved by the two-step process. Triangular voids surrounded by the {10-11} GaN, $-c$-plane GaN, and {22-43} sapphire surface were formed. The GaN-layer surfaces consisted of *m*-plane and {10-11} facets. The width of the GaN-layer growth region toward the $-c$-direction was approximately 100 nm.

Next, we investigated the temperature dependence of the growth rate for the second step. Figure 3.4a, b show cross-sectional SEM images of the GaN layers grown at 900 and 925 °C, respectively, in the second stage. Both GaN layers were perfectly coalesced. However, for the growth temperature of 925 °C, a large GaN growth region toward the $-c$-direction was observed. The width of this region was greater than 1,000 nm. We found that the optimization of growth temperature was essential to form a continuous GaN layer and to suppress GaN growth toward the—*c*-direction.

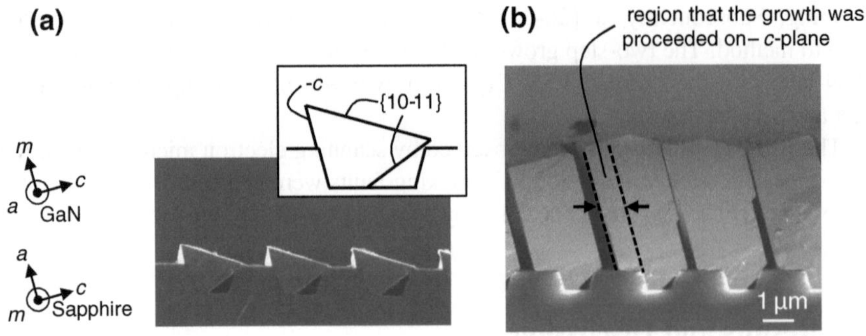

Fig. 3.2 Cross-sectional SEM images of semipolar {20-21} GaN layers grown (**a**) in initial stage and (**b**) for 5 h by one-step growth

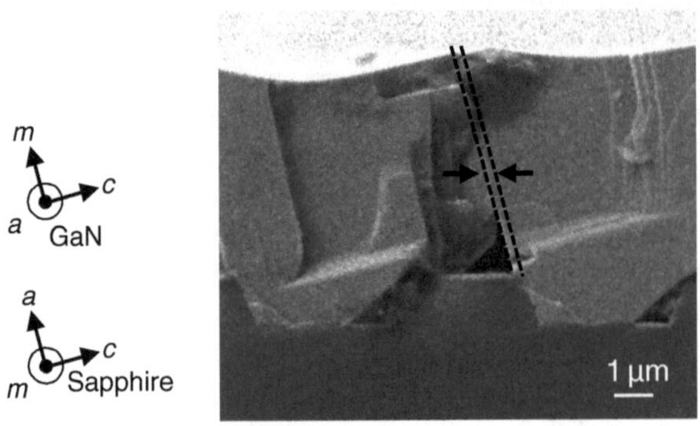

Fig. 3.3 Cross-sectional SEM image of semipolar {20-21} GaN layer grown by two-step method. Growth temperatures of first step/second step were 1,000 °C/900 °C. Dashed lines indicate the region of GaN-layer growth toward the—*c*-direction

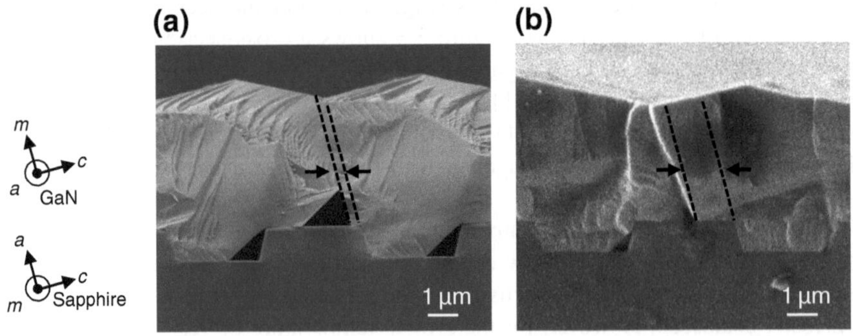

Fig. 3.4 Cross-sectional SEM images of semipolar {20-21} GaN layers grown by using two-step growth method. The growth temperatures of the first step/second step were (**a**) 1,025 °C/900 °C and (**b**) 1,025 °C/925 °C. Dashed lines indicate the region of GaN growth toward the—*c*-direction

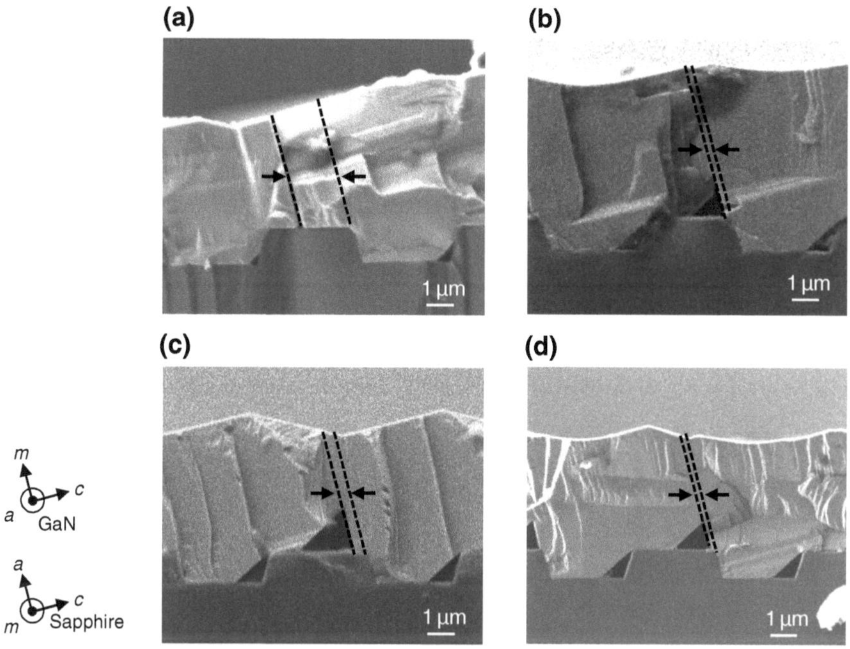

Fig. 3.5 Cross-sectional SEM images of semipolar {20-21} GaN layers grown by two-step growth method. The V/III ratios were (**a**) 147 (**b**) 294 (**c**) 882, and (**d**) 2,942. The growth temperatures of the first step/second step were fixed at 1,000 °C/900 °C, respectively. Dashed lines indicate the regions of GaN-layer growth toward the—*c*-direction

The choice of the V/III ratio also affects the growth mode and crystalline quality of GaN. Because the GaN surface along the $+c$-plane is terminated by Ga atoms (Ga-polar) and that along the $-c$-plane is terminated by N atoms (N-polar), the growth of GaN along the $-c$-plane GaN was suppressed at low values of the V/III ratio [14]. Finally, therefore, we investigated the effect of the V/III ratio on the growth mode of {20-21} GaN. The growth temperatures of the first and second steps were fixed at 1,000 and 900 °C, respectively. The V/III ratio was varied as 147, 294, 882, and 2,942. Figure 3.5 shows the cross-sectional SEM images of the GaN layers grown at different V/III ratio. The width of the GaN-layer growth region toward the $-c$-direction was approximately 200 nm when the V/III ratio was greater than 294. Despite the lowest V/III ratio of 147, the GaN-layer-width toward the $-c$-direction was greater than 1,500 nm. Although the reason underlying this phenomenon is unclear and under investigation, we speculate that the generation of many facet structures such as $-c$, $+c$, m, and {10-11} planes can complicate the growth mode.

Figure 3.6 shows the near-band-edge (NBE) emission intensity and deep-center emission (yellow luminescence) intensity of the grown GaN layers as a function of the V/III ratio when measured at 4 K. The NBE emission was enhanced drastically with increase in the V/III ratio. The intensity of NBE emission of the sample

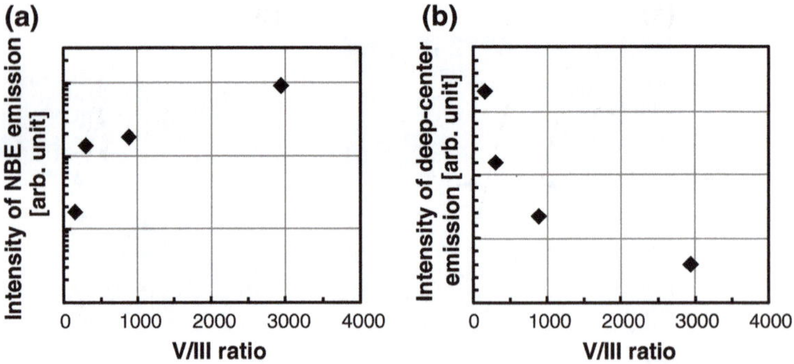

Fig. 3.6 PL intensity of (**a**) NBE emission and (**b**) deep-center emission as function of V/III ratio

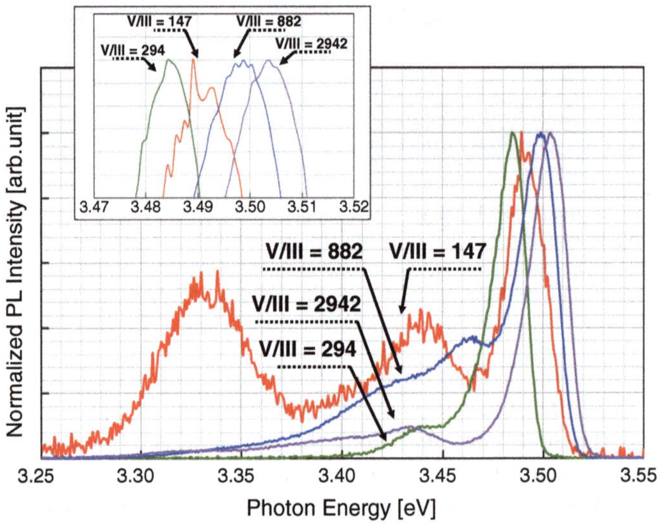

Fig. 3.7 PL spectra of semipolar {20-21} GaN layers grown at various V/III ratio at 4 K. Inset shows magnification of the spectra around the range of NBE emissions

grown with the V/III ratio of 2,942 was approximately 50 times greater than that grown at a V/III ratio of 147. The deep-center emission decreased with increase in the V/III ratio, as shown in Fig. 3.6b. We speculate that N vacancies are suppressed at large V/III ratio.

Figure 3.7 shows the PL spectra of semipolar {20-21} GaN layers grown at various V/III ratios measured at 4 K. A peak at around 3.49 eV representing NBE emission was observed for the {20-21} GaN layers grown on the {22-43} PSS. BSFs and prismatic stacking faults (PSFs) were responsible for emission at energies of approximately 3.44 and 3.33 eV, respectively [17]. The I_1 type of BSFs with the lowest formation energies [18] were observed in all samples. The GaN layer with strong emission corresponding to BSFs exhibited a large GaN-layer growth region toward the −*c*-direction. When the V/III ratio was 147, a strong emission corresponding to PSFs was observed. The NBE emission peaks exhibited a gradual shift to the high-energy side with increasing V/III ratio, as shown in inset of Fig. 3.7. The slight difference was attributed to the variation in compressive stress in the sample structures. A large V/III ratio was necessary to grow GaN layers with a reduced concentration of deep-center defects, such as N vacancies, and without lattice relaxation. High-quality template is essential for the high-efficiency devices. We expect that semipolar GaN template with low defect density realize more efficient and cheap as light sources for optoelectronic devices.

3.4 Summary

In conclusion, we achieved the heteroepitaxial growth of semipolar {20-21} GaN on {22-43} PSS by using MOVPE, and we investigated the GaN growth mode of these samples. We succeeded in suppressing growth toward the −*c*-direction at growth temperatures of 1,000 and 900 °C for the first step and the second step, respectively; this resulted in a large reduction in the formation of stacking faults. The PL measurements showed an increase in NBE emission and a decrease in deep-center emission when the two-step growth was performed at higher V/III ratio. We believe that our study will contribute to realize more efficient and cheap light sources for optoelectronic devices.

Acknowledgments This work was partly supported by the “Regional Innovation Cluster Program” (Global Type), and a Grant-in-Aid for Young Scientists B # 24,760,012 funded by the Ministry of Education, Culture, Sports, Science and Technology, Japan

References

1. Funato M, Ueda M, Kawakami Y, Nakamura Y, Kosugi T, Takahashi M, Mukai T (2006) Blue, green, and amber InGaN/GaN light-emitting diodes on semipolar {11-22} GaN bulk substrates. Jpn J Appl Phys 45:L659
2. Enya Y, Yoshizumi Y, Kyono T, Akita K, Ueno M, Adachi M, Sumitomo T, Tokuyama S, Ikegami T, Katayama K, Nakamura T (2009) 531 nm green lasing of InGaN based laser diodes on semi-polar {20-21} free-standing GaN substrates. Appl Phys Express 2:082101
3. Zhao Y, Tanaka S, Pan CC, Fujito K, Feezell D, Speck JS, DenBaars SP, Nakamura S (2011) High-power blue-violet semipolar (20-2-1) InGaN/GaN light-emitting diodes with low efficiency droop at 200 A/cm^2. Appl Phys Express 4:082104
4. Sasaki T, Zembutsu S (1987) Substrateorientation dependence of GaN single-crystal films grown by metalorganic vapor-phase epitaxy. J Appl Phys 61:2533
5. Waltereit P, Brandt O, Trampert A, Grahn HT, Menniger J, Ramsteiner M., Reiche M, Ploog KH (2000) Nitride semiconductors free of electrostatic fields for efficient white light-emitting diodes. Nature 406:865 (London)

6. Armitage R, Hirayama H (2008) M-plane GaN grown on m-sapphire by metalorganic vapor phase epitaxy. Appl Phys Lett 92:092121
7. Baker TJ, Haskell BA, Wu F, Speck JS, Nakamura S (2006) Characterization of planar semipolar gallium nitride films on sapphire substrates. Jpn J Appl Phys 45:L154
8. Baker TJ, Haskell BA, Wu F, Fini PT, Speck JS, Nakamura S (2005) Characterization of planar semipolar gallium nitride films on spinel substrates. Jpn J Appl Phys 44:L920
9. Okada N, Oshita H, Yamane K, Tadatomo K (2011) High-quality {20-21} GaN layers on patterned sapphire substrate with wide-terrace. Appl Phys Lett 99:242103
10. Yamane K, Ueno M, Uchida K, Furuya H, Okada N, Tadatomo K (2012) Reduction in dislocation density of semipolar GaN layers on patterned sapphire substrates by hydride vapor phase epitaxy. Appl Phys Express 5:095503
11. Schwaiger S, Argut I, Wunderer T, Rosch R, Lipski F, Biskupek J, Kaiser U, Scholz F (2010) Planar semipolar(10-11) GaN on (11-23) sapphire. Appl Phys Lett 96:231905
12. Honda Y, Kawaguchi Y, Ohtake Y, Tanaka S, Yamaguchi M, Sawaki N (2001) Selective area growth of GaN microstructures on patterned (111) and (001) Si substrates. J Cryst Growth 230:346
13. Okada N, Kurisu A, Murakami K, Tadatomo K (2009) Growth of Semipolar (11–22) GaN layer by controlling anisotropic growth rates in r-plane patterned sapphire substrate. Appl Phys Express 2:091001
14. Yamane K, Inagaki T, Hashimoto Y, Koyama M, Okada N, Tadatomo K (2014) Characterization of structural defects in semipolar {20-21} GaN layers grown on {22-43} patterned sapphire substrates. Jpn J Appl Phys 53:035502
15. Kawashima T, Nagai T, Iida D, Miura A, Okadome Y, Tsuchiya Y, Iwaya M, Kamiyama S, Amano H, Akasaki I (2007) Reduction in defect density over whole area of (1-100) m-plane GaN using one-sidewall seeded epitaxial lateral overgrowth. Phys Status Solidi B 244:1848
16. Miyake H, Motogaito A, Hiramatsu K (1999) Effects of reactor pressure on epitaxial lateral overgrowth of GaN via low-pressure metalorganic vapor phase epitaxy. Jpn J Appl Phys 38:L1000
17. Mei J, Srinivasan S, Liu R, Ponce FA, Narukawa Y, Mukai T (2006) Prismatic stacking faults in epitaxially laterally overgrown GaN. Appl Phys Lett 88:141912
18. Stampfl C, Van de Walle CG (1998) Energetics and electronic structure of stacking faults in AlN, GaN, and InN. Phys Rev B 57:R15052

Chapter 4
Development of a Low-Noise Three-Dimensional Imaging LIDAR System Using Two 1×8 Geiger-Mode Avalanche Photodiode Arrays

Sung Eun Jo, Tae Hoon Kim and Hong Jin Kong

Abstract In this paper, a three-dimensional imaging LIDAR system using two 1×8 Geiger-mode avalanche photodiode (GmAPD) arrays is presented. A passively Q-switched microchip laser is used as a light source and a compact PCI (cPCI) system, which includes a time-to-digital converter (TDC), is set up for fast signal processing. Clear 3D images with a fast acquisition speed are obtained by the proposed LIDAR system with using two 1×8 GmAPD arrays for the reduction of false alarms at the TOF data acquisition stage. The software for the three-dimensional visualization is developed for the system of a 1×8 GmAPD array. The range resolution of the system is measured at 100 m, and 3D images acquired by one GmAPD array and two GmAPD arrays during the daytime.

Keywords LIDAR • LADAR • Laser radar • Geiger-mode avalanche photodiodes • Noise • 3D image LIDAR system

S. E. Jo
Department of Aerospace Engineering, KAIST, 373-1 Guseong-dong, Yuseong-gu, Daejeon 305-701, South Korea
e-mail: tjddms98@kaist.ac.kr

T. H. Kim
Image Development Team, System LSI Division, Semiconductor Business, Samsung Electronics Co. Ltd., Suwon 466-711, South Korea
e-mail: kimth@kaist.ac.kr

H. J. Kong (✉)
Department of Physics, KAIST, 373-1 Guseong-dong, Yuseong-gu, Daejeon 305-701, South Korea
e-mail: hjkong@kaist.ac.kr

R. Tutsch et al. (eds.), *Progress in Optomechatronic Technologies*,
Lecture Notes in Electrical Engineering 306, DOI: 10.1007/978-3-319-05711-8_4,

4.1 Introduction

Light detection and ranging (LIDAR) has recently become an important method for both distance measurements and the acquisition of 3D images. Many research groups have used a Geiger-mode avalanche photodiode (GmAPD) as a detector in the 3D imaging LIDAR system due to its extremely high detection sensitivity and a simple readout integrated circuit [1–6]. However, a GmAPD has some disadvantages when used in a 3D imaging LIDAR system. First, the dark counts arising by thermal noise in the depletion region generate false alarms during the stage of the signal processing. Second, a GmAPD is independent of the optical intensity indicating that it cannot distinguish between a signal and noise. Therefore, a noise removal process is essential in a LIDAR system using a GmAPD for clear 3D images.

Heriot-Watt University is the one of the main research centers using a GmAPD as a detector. The method of time-correlated single-photon counting (TCSPC) is used to acquire the distance to the object. It estimates the object distance by thresholding data at a fixed level of the peak height in a TCSPC histogram, but it is time-consuming to repeat the measurements many times to obtain the TCSPC histogram (typically 10^4~10^6) [5]. MIT Lincoln Laboratory has also developed a 3D imaging LIDAR system with a GmAPD. Their 3D imaging LIDAR system is a compact, light-weight system that shows good performance. However, it creates much noise, necessitating much time to remove the noise for clear 3D images as this is done with a series of image processing algorithms [7].

For clear 3D images, the stage of the removal of noise is indispensable. Therefore, we developed a low-noise 3D imaging LIDAR system that can obtain clear 3D image with fast acquisition speed via the reduction of false alarms at the stage of the acquisition of raw time-of-flight (TOF) data with few measurements with two 1×8 GmAPD arrays.

In Sect. 4.2, both the hardware and the software of the low noise 3D imaging LIDAR system are described in detail. The range resolution of the system and 3D images acquired by one GmAPD array and two GmAPD arrays during the daytime are shown in Sect. 4.3. Section 4.4 provides some summary remarks.

4.2 Low-Noise Three-Dimensional Imaging LIDAR System

The LIDAR system is divided into two parts: its hardware and its software. Figure 4.1 shows a schematic diagram of the LIDAR system. A laser pulse is emitted from a light source and passes through the optical system. A small portion of the laser pulse is used to generate the start signal and the rest of the laser pulse is irradiated onto the target. The part of the scattered laser pulse and background light in the field of view (FOV) are collected by the receiving optical system. Then, the laser return pulse and background light are intensity-divided in half by a beam splitter and

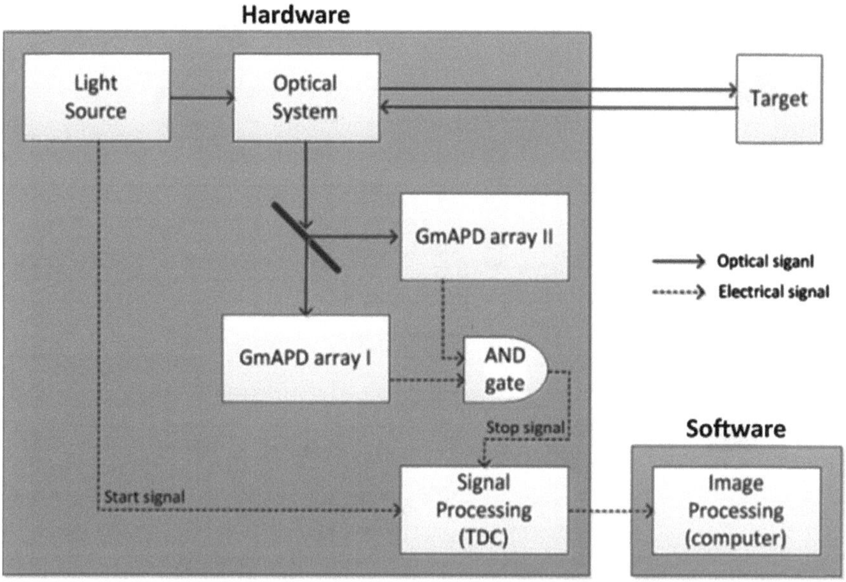

Fig. 4.1 Schematic diagram of the LIDAR system

routed to two GmAPD arrays. An AND gate compares the arrival time of the two GmAPD arrays and a stop signal is generated only if the time difference between the two signals from the GmAPD arrays is less than a fixed value. Although the signal is decreased due to the division into half of the energy of the laser-return pulse, the number of false alarms is decreased drastically because the noise distributed randomly in the time domain is filtered out [8]. TOF data, which is the time difference between the start signal and the stop signal, is generated at the TDC during the signal processing step. The TOF data is then transferred to a distance to visualize a 3D image with the point cloud method in the image processing step.

4.2.1 Hardware

The optical system is shown in Fig. 4.2. A diode-pumped passively Q-switched microchip laser with a second harmonic generation (Alphalas PULSELAS-P-1064-300-FC/SHG) is used as a light source. The start signal cannot be generated by the laser itself, as the laser is passively Q-switched. Therefore, a small portion of laser pulse is transmitted to the photodiode at a 45° reflection mirror. The PD used for generating the start signal is a Thorlab high-speed Si detector. The rest of the laser pulse is reflected by a mirror and collimated by lenses L1 and L2. Due to the single polarization of the laser, a half-wave plate (HWP) is located before the polarization beam splitter (PBS) in order to control both the transmission and

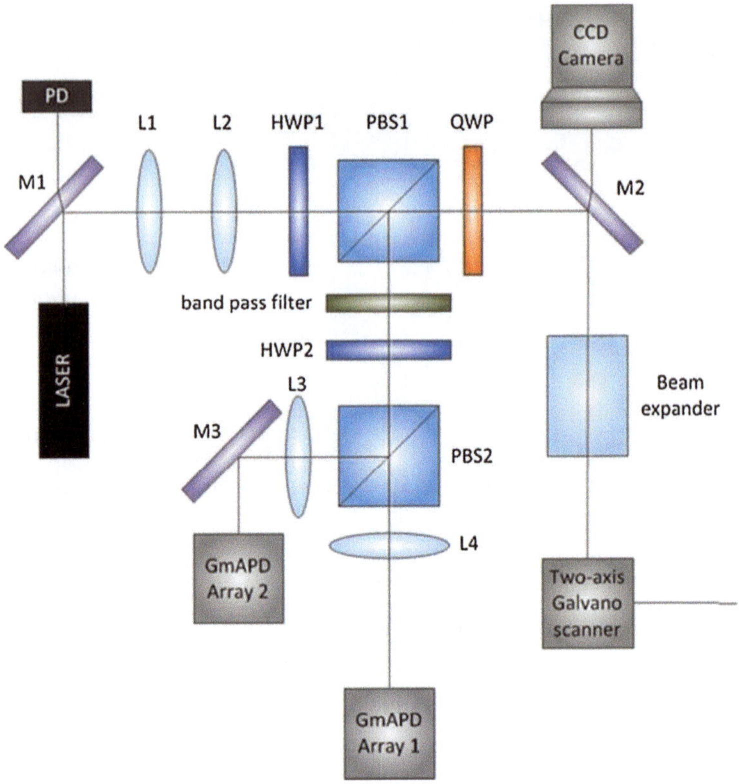

Fig. 4.2 Optical system. (L: lens, HWP: half-wave plate, PBS: polarization beam splitter, QWP: quarter-wave plate, M: mirror, PD: photodiode)

reflection of the laser pulses at PBS1. The transmitted laser pulse passes through a beam expander. The FOV of the system determined by the focal length of lenses L3 and L4 and a beam expander is set to be identical to the laser beam divergence assuming Gaussian propagation. After the beam expander, the laser pulse is guided to the target by a two-axis galvano scanner and is then scattered. The two-axis galvano scanner used for controlling the beam pointing for 1×8 GmAPD array is from Cambridge Technology with a 50 mm aperture mirror. The scattered laser pulse from the target and background light in the FOV of the system are collected into the GmAPD arrays (ID quantique id150-1×8) through the two-axis galvano scanner, a quarter-wave plate (QWP), PBS1, an optical band-pass filter, HWP2, PBS2, and the focusing lenses in that order. Because the microchip laser has a narrow spectral linewidth and temperature stability, the optical band-pass filter, which is centered at a wavelength of 532 nm with a bandwidth (FWHM) of 10 nm and a maximum transmission of 65.91 %, is used for spectral filtering.

A TDC (Agilent U1051A), with six channels and a timing resolution of 50 ps receives both the start and stop signal and measures the time difference between them. The function of the AND gate is to compare the TOFs as measured by GmAPD array 1 and GmAPD array 2. When applying the functionality of the AND gate to the TOFs, calibration is necessary due to the different time-delay characteristics between each pixel in GmAPD arrays 1 and 2. A time bin is defined as a unit that indicates the time interval obtained by dividing the total measurement time by a specific value. The time bin is set to 3 ns, determined by calculating the overall timing jitter of the system.

A charge-coupled device (CCD) camera (Pixelink PL-B953U) with a fixed focal length lens is used as a boresight camera mounted in the optical axis of the system in order to photograph the targets. The FOV of the CCD camera comparatively accords with the FOV of the system.

4.2.2 Software

The algorithm executes a series of functional steps to convert the TOF data obtained by the system into a 3D image. After receiving the laser pulses, the system provides a two-dimensional depth image that includes angle-angle-depth data for software processing and 3D image visualization data. First the TOF data are converted into Cartesian locations. XYZ points at the Cartesian coordinate are calculated using the scanning angle and range information of each pixel. Using the XYZ points, 3D image visualization can be carried out by a point cloud scheme. In the point cloud scheme, each of the XYZ points in the received data is simply plotted in a 3D Cartesian coordinate. Next, there are various algorithms that serve to remove the noise, but these algorithms are unnecessary in the proposed system because the noise was removed during the acquisition of the raw TOF data.

4.3 Range Resolution and 3D Images

The range resolution of a LIDAR system is its ability to distinguish between targets which are very close in either range or bearing. To obtain the range resolution of the system at 100 m, the standard deviations of the TOFs of the proposed system were measured. As shown in Fig. 4.3, the measurement target was located between 99.8 m and 100.2 m, at 10 cm intervals on a motorized translational stage. The standard deviations were obtained with 10,000 laser pulses at each position.

Figure 4.4 shows the range resolution of a pixel of GmAPD array 1. As Fig. 4.4 represents, the location of the target can be distinguished between 100 m and 100.1 m. For a pixel of GmAPD array 1, the 2σ value was 7.4 cm. The average range resolution of all pixels of the two GmAPD arrays (2σ) was 7.6 cm at 100 m. An error is defined as a difference between a true value and a measured value. The average error of all pixels of the two GmAPD arrays is 5.8 mm which is less than the timing resolution of TDC (7.5 mm).

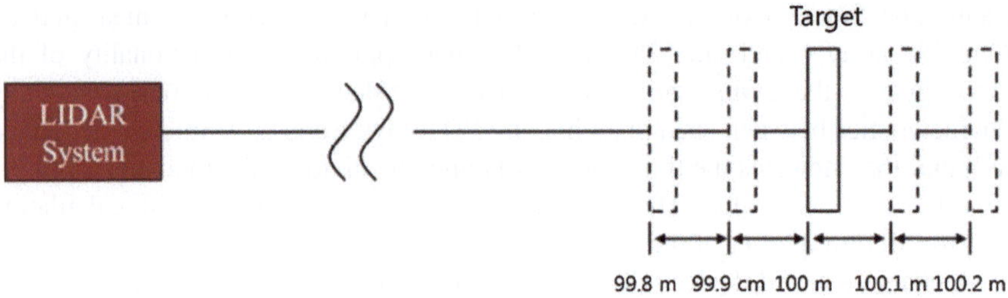

Fig. 4.3 Measurement targets

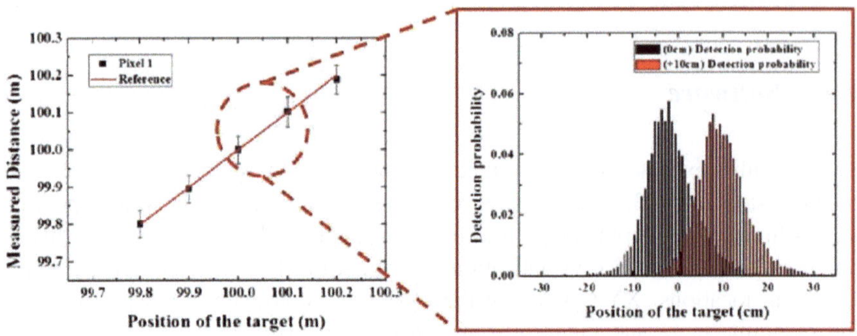

Fig. 4.4 Range resolution of a pixel of GmAPD array 1

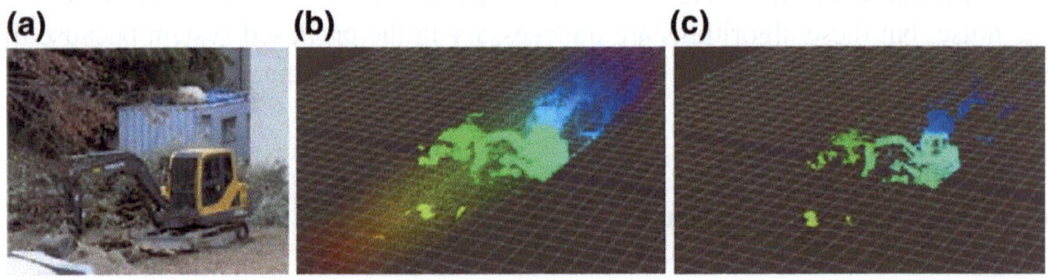

Fig. 4.5 **a** 2D image **b** 3D image acquired by one GmAPD array, and **c** 3D image acquired by two GmAPD arrays

Figure 4.5a shows a two-dimensional image of the scene. The distance of the scene was approximately 60 m. The average rate functions of the noise, defined as the sum of the rate of background photons impinging on the detector and the dark count rate of the GmAPD arrays, were 16 kHz and 20 kHz for GmAPD arrays 1 and 2, respectively. The three-dimensional images were acquired with 1164 points × 1170 points scanning with a field of regard (FOR) of 4°×4°. Figures 4.5b, c show 3D images acquired by one GmAPD array and two GmAPD arrays, respectively. The capacity for a 3D image file acquired by one GmAPD array was 162 MB, and the capacity for a 3D image file acquired by two GmAPD arrays was 3.6 MB.

4.4 Summary

This paper describes a low-noise three-dimensional imaging LIDAR system using two 1×8 GmAPD arrays. A passively Q-switched microchip laser is used as a light source and a cPCI system, which includes a TDC, and is set up for fast signal processing. The proposed LIDAR system is capable of obtaining clear 3D images with a rapid acquisition speed which is achieved by the reduction of false alarms at the TOF data acquisition stage in conjunction with the use of a 1×8 GmAPD array. The software for the three-dimensional visualization is developed for the system of the 1×8 GmAPD array. The range resolution of the system was measured at 100 m and 3D images acquired by one GmAPD array and by two GmAPD arrays during the daytime are presented.

Acknowledgements This work was supported by "Dual Use Technology Program" at Agency for Defense Development (ADD) of the Republic of Korea((UM12012RD1).

References

1. Albota MA, Aull BF, Fouche DG, Geinriches RM, Kocher DG, Marino RM, Moony JG, Newbury NR, O'Brien ME, Player BE, Willard BC, Zayhowski JJ (2002) Three-dimensional laser radar with geiger-mode avalanche photodiode arrays. MIT Linc J 13:2
2. Aull BF, Loomis AH, Young DJ, Heinrichs RM, Felton BH, Daniels PJ, Landers DJ (2002) Geiger-mode avalanche photodiodes for three-dimensional imaging. Linc Lab J 13:2
3. Marino RM, Davis WR (2005) Jigsaw: a foliage-penetrating 3D imaging laser radar system. Linc Lab J 15:23
4. Massa JS, Wallace AM, Buller GS, Fancey SJ, Walker AC (1997) Laser depth measurement based on time-correlated single-photon counting. Opt Lett 22:543
5. Massa J, Buller G, Walker A, Smith G, Cova S, Umasuthan M, Wallace A (2002) Optical design and evaluation of a three-dimensional imaging and ranging system based on time-correlated single-photon counting. Apple Opt 41(6):1063
6. McCarthy A, Collins RJ, Krichel NJ, Fernandes V, Wallace AM, Buller GS (2009) Long-range time of flight scanning sensor based on high speed time-correlated photon counting. Apple Opt 48(32):6241
7. Cho P, Anderson H, Hatch R, Ramaswami P (2006) Real-time 3D ladar imaging. Linc Lab J 16:147
8. Kong HJ, Kim TH, Jo SE, Oh MS (2011) Smart three-dimensional imaging ladar using two Geiger-mode avalanche photodiodes. Opt Express 19(20):19323

Chapter 5
Quick Light-Mixing for Image Inspection Using Simplex Search and Robust Parameter Design

HyungTae Kim, KyeongYong Cho, SeungTaek Kim, Jongseok Kim and Sungbok Kang

Abstract Finding the optimal illumination conditions for industrial machine vision is iterative and time-consuming work. This study discusses simplex search for automatic illumination and the Taguchi method for algorithm tuning. The simplex search could find the illumination inputs to obtain a fine image, but the degree of fineness and processing time varied according to the algorithm parameters. The mechanism from the inputs and the degree of fineness were complex and nonlinear, so it was hard to find the best parameters through conventional experiments. To address these issues, the Taguchi method was applied to the simplex search and used to figure out the best parameter combination with a minimum number of experiments. The targets of Taguchi analysis were finer images and fewer iterations. An $L_{25}(5^5)$ orthogonal array was constructed for 5 parameters and 5 levels, and was filled with the experimental results. A new combination of parameters determined by Taguchi analysis was applied to retests. The retest results showed fewer iterations with a fine image that was close to the best case. The Taguchi method can reduce the amount of work required to set parameters for optimal illumination.

Keywords Color mixer • Optimal illumination • Simplex search • Robust design of experiments • Taguchi method

H. Kim (✉) · S. Kim · J. Kim · S. Kang
Manufacturing System R&D Group, KITECH, 89, YangDae-GiRo, ChenAn 331-825, ChungNam, South Korea
e-mail: htkim@kitech.re.kr
URL: http://www.kitech.re.kr

K. Cho
UTRC, KAIST, 291, DaeHak, YuSeong, DaeJeon 305-701, South Korea
e-mail: yong00@kaist.ac.kr

R. Tutsch et al. (eds.), *Progress in Optomechatronic Technologies*,
Lecture Notes in Electrical Engineering 306, DOI: 10.1007/978-3-319-05711-8_5,

5.1 Introduction

Auto focusing and auto lighting have become popular for acquiring fine images in industrial vision. Studies on auto focusing have been common, and have usually issued on contrast and sharpness [1, 2]. An auto focusing system adjusts the distance between the target pattern and a camera and simultaneously obtains image properties for focusing [3]. In contrast, auto lighting has not been addressed as much as in previous studies, and examples have involved projectors [4] and multiplex illumination [5]. Auto lighting for a single color is applied to some commercial machines. Quick search algorithms were considered for industrial machine vision in single-color system [6]. Image acquisition is a procedure that involves illumination, reflection on a target pattern, spectral responses of optics, digitalizing images, and numerical processing. A few cases have tried to describe the procedure using spectral responses, but the procedure is nonlinear, complex, and unpredictable in general [7].

Synthesized light using a color mixer can improve the image quality in industrial machine vision. Light mixing is complex and difficult because multiple sources are handles and multiple optical devices are combined. Light mixing mechanisms were reported in some studies, such as Muthu's RGB light engine [8], Sun's tunable lamp [9], Esparza's light pipe [10], Muschaweck's micro-lens array [11] and Gorkom's Collimator [12]. These studies focused on the light mixing mechanisms and the exact color generation. However, they did not discuss applications to machine vision or efficient use of the mechanism. Cases were reported on optimal color mixing for machine vision. Lee proposed a light mixing method for an endoscope based on the optical spectrum [13]. Histogram-based [14] and optimum-based algorithms [15] were also proposed for finding optimal conditions of a color mixer. Optimum-based algorithms were tested for optimal illumination, and showed potential for shortening search time.

However, it is a laborious task to set the color mixer and the algorithm for industrial vision systems. The relation between illumination and fineness is nonlinear and hard to describe, so it is difficult to tune the parameters of search algorithms. Therefore, this study proposes how to set and tune the parameters of simplex search for optimal color illumination with a minimum number of experiments. The Taguchi method is one of the experimental design methods used for this goal. This method was applied to reduce the number of iterations needed for obtaining fine images that are close to the best case.

5.2 Simplex Search for Color Mixer

5.2.1 Defining a Cost Function

Optimal illumination is a problem to find the maximum sharpness by adjusting multiple light sources in a color mixer. The equations of the optimum problem were reported in previous research [15]. The N voltage inputs of the multiple sources are defined as follows.

$$V = (v_1, v_2, \ldots, v_N) \tag{5.1}$$

The voltage inputs are changed into synthesized light through a color mixer. The synthesized light is reflected on the surface of a target object, and the reflected light arrives at a vision camera. The optical response of a pixel in the camera becomes the grey levels, which indicate the intensity of the light. The image acquired from the camera is the assembly of all the pixels. The grey levels of the pixel in the image are used to calculate the sharpness; the degree of fineness. The sharpness; σ is affected by the voltage input V, and their relation can be defined as an arbitrary function f. The optimum methods are derived to find the minimum, and the cost function is defined as follows:

$$\mathit{find\,min} \quad \rho = -\sigma = -f(V) \quad \mathit{for} \quad \forall V \tag{5.2}$$

5.2.2 Simplex Search

The simplex search is briefly explained in this chapter. A simplex, or a probe network, is organized using N + 1 probing points for N inputs. The color mixer has three inputs of RGB, so the simplex has N = 3 inputs and a tetrahedron shape. The probe network has N + 1 = 4 probing points. The probe network obtains values from the probing points and sorts the evaluations [16]. The probe network of this case can be defined as follows.

$$\begin{bmatrix} W_1 \\ W_2 \\ W_3 \\ W_4 \end{bmatrix} = \begin{bmatrix} v_{I,1} & v_{II,1} & v_{III,1} \\ v_{I,2} & v_{II,2} & v_{III,2} \\ v_{I,3} & v_{II,3} & v_{III,3} \\ v_{I,4} & v_{II,4} & v_{III,4} \end{bmatrix} \tag{5.3}$$

$$\begin{bmatrix} \rho_1 \\ \rho_2 \\ \rho_3 \\ \rho_4 \end{bmatrix} = \begin{bmatrix} -f(W_1) \\ -f(W_2) \\ -f(W_3) \\ -f(W_4) \end{bmatrix} \tag{5.4}$$

The sequence of simplex search is composed of expansion, contraction and shrinkage, which are determined by testing points. The sequence is iterated until its size becomes smaller than a terminal condition. The terminal condition usually has a small value ε, and the following is the most common:

$$\begin{aligned} |W_1 - W_4| &< \varepsilon \\ |\rho_1 - \rho_4| &< \varepsilon \end{aligned} \tag{5.5}$$

A middle point is defined for an extension line from the worst point.

$$W_{mid} = \frac{1}{N}\sum_{i=2}^{4} W_i \tag{5.6}$$

$$\begin{aligned} d &= W_{mid} - W_1 \\ W_{line} &= W_{mid} + \lambda_{line} d \end{aligned} \tag{5.7}$$

The first test point is defined on the extension line at a distance *d* from the middle point.

$$W_{t1} = W_{mid} + d \tag{5.8}$$

If ρ_{t1} is a new minimum, a second test point is defined outside of the reflection point. The worst point is replaced with the expansion point.

$$W_{t2} = W_{mid} + \lambda_e d \tag{5.9}$$

If ρ_{t1} is not a new minimum, a second test point is defined in the simplex on the extension line. If ρ_{t2} is a new minimum, the worst point is replaced with the contraction point.

$$W_{t2} = W_{mid} - \lambda_c d \tag{5.10}$$

If the ρ_{t2} is not a new minimum, the simplex is shrunken towards the current minimum, the best point W_4.

$$W_{i,new} = (W_4 + W_i)/2 \quad (i = 1, 2, 3) \tag{5.11}$$

The average sharpness, $\bar{\sigma}$, is the representative value in the probe network.

$$\bar{\sigma} = \sum_{i}^{4} \sigma_i \tag{5.12}$$

5.3 Robust Parameter Design

5.3.1 Background

The Taguchi method was applied to design a machine vision system and the main factors of the system capabilities were determined using ANOVA [17]. A Taguchi-based method was also applied to tune the parameters of the image processing algorithms. Su developed an inspection algorithm based on neural networks for defects in fabrics [18]. There have also been cases of algorithm optimization for face recognition and for the analysis of human gestures [19, 20]. Most studies on illumination using the Taguchi method were designed of optimal light for applications; such as lithography and LCD backlights [21, 22]. The parameters of the Taguchi method are controllable factors that affect the performances of a target system. Evaluation of Taguchi method can be classified in the following three categories. "Nominal the best" is the case of a specified target [23].

$$SN = 20 \log \left(\frac{\bar{y}}{S^2} \right) \tag{5.13}$$

"Larger the better" is the case in which large characteristics are require.

$$SN = -10 \log \left(\frac{1}{n} \sum_{i=1}^{n} \frac{1}{y_i^2} \right) \tag{5.14}$$

"Smaller the better" is the inverse case of "larger the better".

$$SN = -10 \log \left(\frac{1}{n} \sum_{i=1}^{n} y_i^2 \right) \tag{5.15}$$

The simplex search shortens the time to find the optimal illumination, but it is necessary to tune the parameters of the search algorithm to increase the performance and reliability. Previous research introduced simplex search for optimal illumination but only the possibility of quick convergence without any parameter optimization was shown [15]. The parameters of simplex search are the initial voltage inputs, the expansion scale, and the contraction scale. The reflection and shrinkage scales were set with values which were typically used. Many experiments are needed to tune the five factors for effective and reliable convergence of the simplex search, so robust parameter design is essential to minimize the number of experiments [24].

5.3.2 *Test Apparatus*

A color mixer and the machine vision were constructed in previous research [15]. The color mixer has RGB sources, and lights were mixed in a mixing chamber connected to each source through an optical fiber. The mixed light illuminates a target pattern through a coaxial lens. The inputs of the sources are adjusted by a probe network. The sharpness is calculated after image acquisition at each probing point. The voltage inputs of the light sources are adjusted by the simplex method. The simplex search is iterated until the terminal condition is satisfied. The target patterns were EP910JC35 (ALTERA) and Z86E3012KSE (ZILOG) commercial ICs, as shown in Fig. 5.1.

5.3.3 *Experiment Design for Simplex Search*

The simplex for RGB inputs is defined as an N = 3 problem and there are 12 initial inputs since there are N = 3 × 4 probing points. This can make the

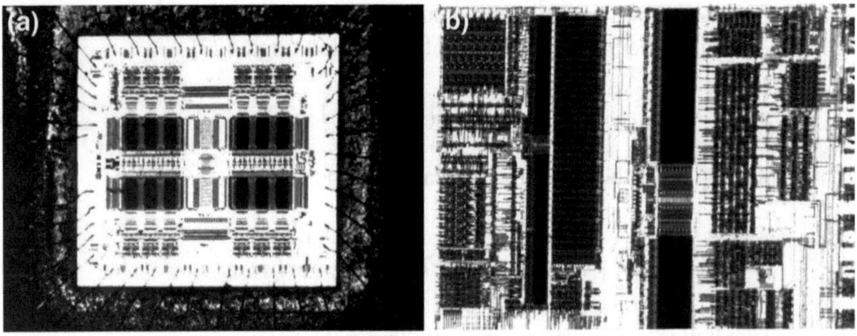

Fig. 5.1 Target patterns acquired by optimal illumination using simplex search (**a**) EP910JC35 (ALTERA) (**b**) Z86E3012KSES (ZILOG)

Table. 5.1 Control factors and levels for simplex search

Factors	Code	Level				
		1	2	3	4	5
V_I: initial V_R	A	0.5	1.0	1.5	2.0	2.5
V_{II}: initial V_G	B	0.5	1.0	1.5	2.0	2.5
V_{III}: initial V_B	C	0.5	1.0	1.5	2.0	2.5
λ_e: expansion scale	D	1.5	2.0	2.5	3.0	3.5
λ_c: contraction scale	E	0.1	0.3	0.5	0.7	0.9

parameter design complex, so the initial simplex is formed by three values, interconnecting the origin and diagonal corners. The corners can be assigned using $V_{I,0}$, $V_{II,0}$, and $V_{III,0}$. The initial probe network is represented by the following equation;

$$\begin{bmatrix} W_1 \\ W_2 \\ W_3 \\ W_4 \end{bmatrix} = \begin{bmatrix} 0 & 0 & 0 \\ V_{I,0} & V_{II,0} & 0 \\ 0 & V_{II,0} & V_{III,0} \\ V_{I,0} & 0 & V_{III,0} \end{bmatrix} \quad (5.16)$$

The control factors for the RGB mixers are $V_{I,0}$, $V_{II,0}$, $V_{III,0}$, λ_e, and λ_c. The purpose of the tuning is to decrease the number of iterations and increase ρ_4. The control factors and levels for the simplex search in the color mixer are summarized in Table 5.1. A–E in the table are symbols of the control factors, and there are 5 levels. Total number of experiments required is $5^5 = 3125$, which requires a long time to complete. Thus, the Taguchi method is used to reduce the number of experiments, and can be defined as an orthogonal array of $L_{25}(5^5)$, as shown in the Table 5.2. The analysis of the Taguchi method was focused on increasing the maximum sharpness and decreasing the number of iterations.

Table 5.2 Orthogonal array of $L_{25}(5^5)$ for pattern A and B

Run #	Control factors					Pattern A					Pattern A				
	A	B	C	D	E	$\bar{\sigma}_{max}$	k	V_R	V_G	V_B	$\bar{\sigma}_{max}$	k	V_R	V_G	V_B
1	1	1	1	1	1	390.69	18	0.004	0.003	1.296	358.18	17	0.951	0.006	0.002
2	1	2	2	2	2	389.33	29	0.915	0.016	0.475	358.29	22	0.968	0.003	0.000
3	1	3	3	3	3	389.26	31	0.865	0.000	0.490	358.33	26	0.974	0.003	0.000
4	1	4	4	4	4	389.33	43	0.898	0.001	0.471	358.22	33	0.963	0.004	0.000
5	1	5	5	5	5	390.19	63	0.015	0.028	1.373	358.28	53	0.971	0.012	0.000
6	2	1	2	3	4	389.30	28	0.910	0.000	0.463	358.15	55	0.965	0.025	0.007
7	2	2	3	4	5	388.59	37	1.528	0.013	0.000	358.25	35	0.978	0.000	0.000
8	2	3	4	5	1	389.13	22	1.057	0.034	0.308	348.99	17	0.767	0.009	0.285
9	2	4	5	1	2	388.58	18	1.557	0.000	0.000	358.19	28	1.004	0.000	0.000
10	2	5	1	2	3	389.19	20	0.926	0.011	0.501	357.59	37	0.932	0.047	0.017
11	3	1	3	5	2	388.77	27	0.596	0.030	0.817	357.94	28	0.987	0.017	0.008
12	3	2	4	1	3	390.54	33	0.000	0.000	1.296	358.19	39	0.965	0.027	0.000
13	3	3	5	2	4	389.24	29	0.982	0.002	0.431	358.33	42	0.955	0.000	0.000
14	3	4	1	3	5	390.49	36	0.000	0.005	1.276	357.66	87	0.946	0.052	0.009
15	3	5	2	4	1	390.48	23	0.000	0.000	1.287	358.29	26	0.964	0.000	0.000
16	4	1	4	2	5	390.43	50	0.001	0.000	1.294	358.18	69	0.977	0.024	0.000
17	4	2	5	3	1	389.14	30	0.933	0.014	0.483	358.25	60	0.973	0.017	0.000
18	4	3	1	4	2	390.39	20	0.000	0.000	1.366	354.69	25	0.623	0.233	0.011
19	4	4	2	5	3	390.16	27	0.007	0.037	1.344	349.89	47	1.037	0.070	0.089
20	4	5	3	1	4	389.90	41	0.019	0.024	1.445	358.20	45	0.960	0.016	0.006
21	5	1	5	4	3	388.66	39	1.268	0.072	0.161	358.23	28	0.935	0.032	0.000
22	5	2	1	5	4	390.22	19	0.022	0.017	1.307	358.28	32	0.948	0.000	0.000
23	5	3	2	1	5	390.45	29	0.000	0.000	1.315	358.24	58	0.953	0.000	0.000
24	5	4	3	2	1	388.63	32	0.496	0.049	0.809	356.56	23	0.788	0.123	0.000
25	5	5	4	3	2	388.53	38	0.804	0.144	0.400	352.35	39	0.559	0.329	0.009

5.4 Results

5.4.1 Experimental Results

25 cases of experiments were carried out for the L_{25} (5^5) orthogonal array. The terminal states of simplex search for patterns A and B are summarized in Table 5.2. The terminal states are $\bar{\sigma}_{max}$, k, V_R, V_G and V_B. The $\bar{\sigma}$ values were almost the same, but the inputs of the terminal states were different. When $\bar{\sigma}$ increased, the voltage inputs were near the results of equal search. The control factors are not very effective on $\bar{\sigma}$, but effective on k. The fluctuation of $\bar{\sigma}$ and k shows that $\bar{\sigma}_{max}$ and k can be adjusted for a desirable direction by the parameter design, which indicates increase of $\bar{\sigma}_{max}$ and decrease of k.

5.4.2 Taguchi Analysis

The object of parameter design is to increase $\bar{\sigma}$ and decrease k for higher search performance, which corresponds to higher sharpness and fewer iterations. So, in the Taguchi analysis, "larger the better" can be applied to $\bar{\sigma}$, and "smaller the better" can be applied to k. The orthogonal array and the experimental results were input into the MINITAB analysis tool.

The plots in Fig. 5.2 show the influence of the control factors for pattern A in simplex search. The variation of $\bar{\sigma}$ is relatively small, and k is more sensitive. So, the parameters were mainly selected by considering k. The governing factors for k were E, C, and A, whose variations were more stiff than with other factors. C and E showed approximately decreasing trends, but the others had minima in the middle. A new combination was determined from the factors which had the largest value in each plot. The new combination was $A_2B_3C_1D_1E_1$ for pattern A. The analysis results for pattern B are shown in the plots of Fig. 5.3. The responses of $\bar{\sigma}$ were smaller than those of k, which showed the same trend as pattern A. A, B, and E showed decreasing trends and the governing factors were A and E. The largest values in the plot were also selected and the new combination for pattern B was $A_1B_1C_1D_4E_2$.

The new combinations were retested in the color mixer, and the results are summarized in Table 5.3. The $\bar{\sigma}$ values obtained by the Taguchi method were almost the same as the best cases in the orthogonal array, but the final voltages were different for pattern A. The k's after Taguchi method were smaller than those of the orthogonal array. k was mainly considered in the Taguchi analysis, so the k was reduced, but $\sigma_{\max}$ was hardly affected. The color mixer and machine vision form a nonlinear system, and there can be many local maximums near a global maximum, so voltage inputs by the Taguchi method were different form the exact solutions. However, the difference in sharpness between the exact solution and

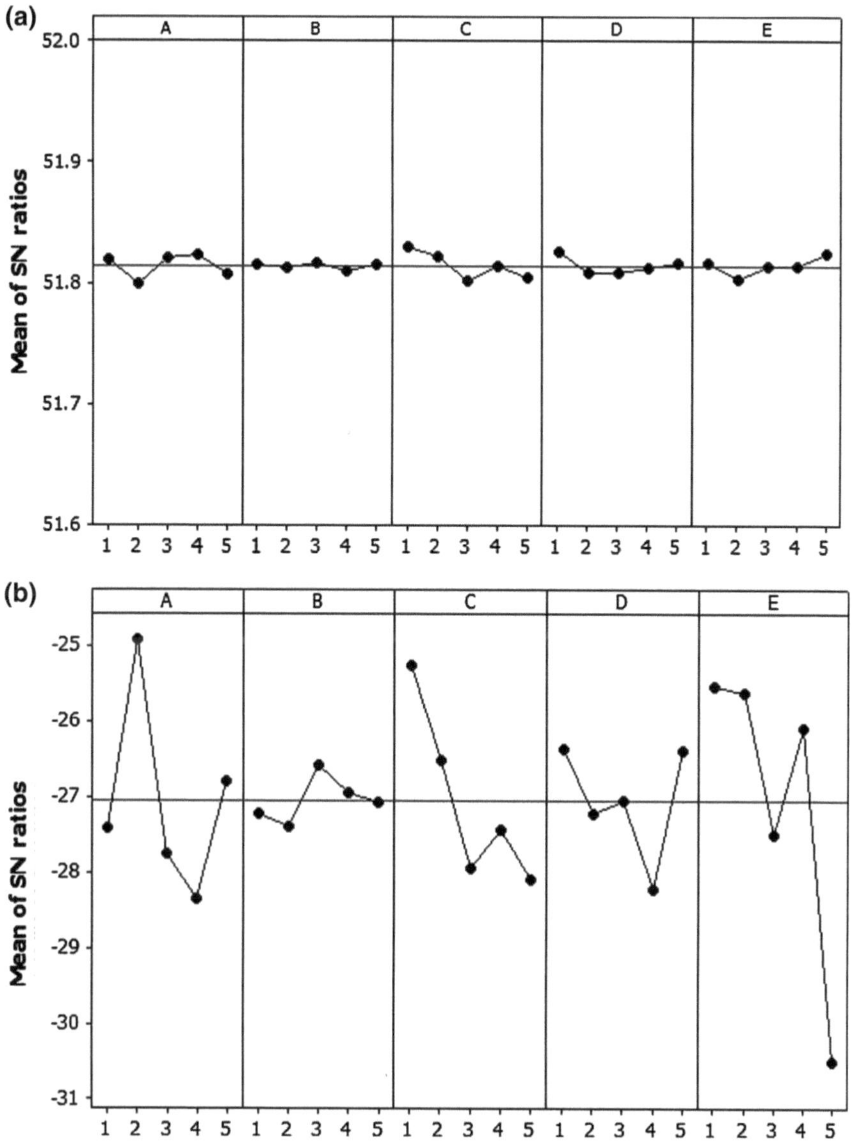

Fig. 5.2 SN ratios of control factors for pattern A. **a** SN ratio for sharpness. **b** SN ratio for the number of iterations

the Taguchi solution was minute, and there were fewer iterations, so the Taguchi method can supply useful parameters for the simplex search without many experiments. The Taguchi method for parameter tuning does not guarantee finding a global maximum, but it can supply useful solution near the global minimum.

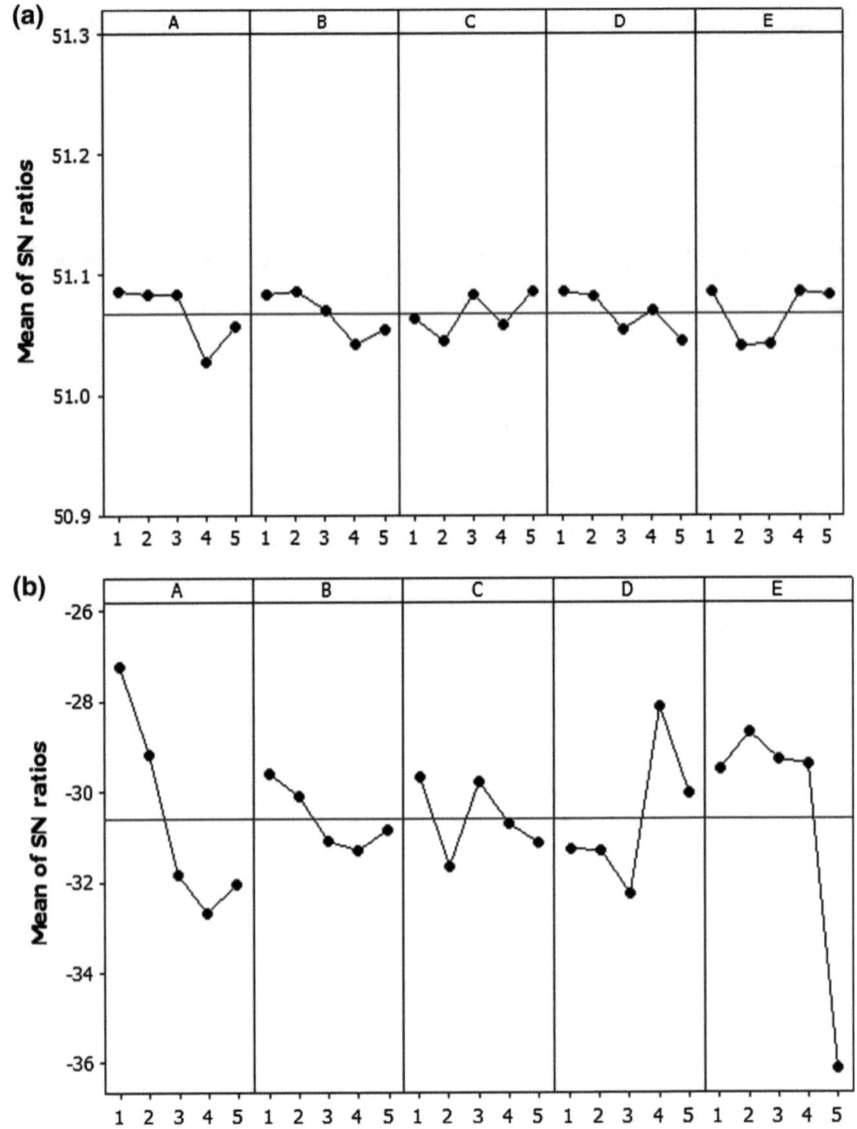

Fig. 5.3 SN ratios of control factors for pattern B. **a** SN ratio for sharpness. **b** SN ratio for the number of iterations

Table 5.3 Retrial results by Taguchi analysis

The best results	Control factors					Values				
	A	B	C	D	E	$\bar{\sigma}_{max}$	k	V_R	V_G	V_B
Pattern A: the orthogonal array	1	1	1	1	1	390.66	14	0.005	0.004	1.307
Pattern B: the orthogonal array	1	1	1	1	1	357.42	12	0.945	0.017	0.013
Pattern A: the equal search	–	–	–	–	–	392.76	–	0.0	0.0	1.2
Pattern B: the equal search	–	–	–	–	–	358.87	–	1.0	0.0	0.0
Pattern A: Taguchi analysis	2	3	1	1	1	391.49	13	1.148	0.000	0.626
Pattern B: Taguchi analysis	1	1	1	4	2	357.70	10	0.964	0.000	0.000

5.5 Conclusion

Simplex search with parameters designed by the Taguchi method has been proposed for the purpose of optimal illumination and efficient use of a color mixer for industrial machine vision. The design parameters of the simplex search were the initial voltages, the expansion scale, and the contraction scale. An L_{25} (5^5) orthogonal array was constructed for the Taguchi analysis. The aim of parameter design was focused on increasing image quality and shortening processing time. A new combination of control factors was obtained after the analysis. The results of retests show that the image quality slightly increases and the number of iterations decreases, as is intended for the Taguchi analysis. The proposed method can contribute to reducing the work needed to set the optimal illumination conditions in industrial machine vision.

Acknowledgments This work was funded and supported by the Korea Institute of Industrial Technology. The authors are grateful to AM Technology (http://www.amtechnology.co.kr) for supplying RGB mixable color sources.

References

1. Sun Y, Duthaler S, Nelson BJ (2004) Autofocusing in computer microscopy: selecting the optimal focus algorithm. Microsc Res Tech 65(3):139–149
2. Jeon J, Yoon I, Kim D, Lee J, Paik J (2010) Fully digital auto-focusing system with automatic focusing region selection and point spread function estimation. IEEE Trans Cons Elec 56(3):1204–1210
3. Kim HT, Kang SB, Kang HS, Cho YJ, Kim JO (2011) Optical distance control for a multi focus image in camera phone module assembly. Int J Precis Eng Manuf 12(5):805–811
4. Varaun AV (2011) Adaptive lighting for machine vision applications. In: Proceedings of IEEE CRV, pp 140–145
5. Schechner YY, Nayar SK, Belhumeur PN (2007) Multiplexing for optimal lighting. IEEE Trans Patt Anan Mach Int 29(8):1339–1354
6. Kim HT, Kim ST, Cho YJ (2012) A review of light intensity control and quick optimum search in machine vision. In: Proceedings on international symposium optomechatronic technologies

7. Arecchi AV, Messadi T, Koshel RJ (2007) Field guide to illumination. SPIE Press, Washington
8. Muthu S, Gaines J (2003) Red, green and blue LED-based white light source: implementation challenges and control design. In: Proceedings on the IEEE industry applications conference, vol 1. pp 515–522
9. Sun CC, Moreno I, Lo YC, Chiu BC, Chien WT (2011) Collimating lamp with well color mixing of red/green/blue LEDs. Opt Express 20(S1):75–84
10. Esparza D Moreno I (2010) Color patterns in a tapered lightpipe with RGB LEDs. In: Proceedings of SPIE, vol 7786
11. Muschaweck J (2011) Randomized micro lens arrays for color mixing. In: Proceedings of SPIE, vol 7954
12. van Gorkom RP, van AS, MA, Verbeek GM, Hoelen CGA, Alferink RG, Mutsaers CA, Cooijmans H (2007) Etendue conserved color mixing. In: Proceedings of SPIE, vol 6670
13. Lee MH, Seo DK, Seo BK, Park JI (2011) Optimal illumination spectrum for endoscope. Korea-Japan Joint Workshop on frontiers of computer vision, pp 1–6
14. Kim HT, Kim ST, Cho YJ (2012) An optical mixer and RGB control for fine images using grey scale distribution. Int J Optomecha 6(3):213–225
15. Kim HT, Kim ST, Kim JS (2013) Mixed-color illumination and quick optimum search for machine vision. Int J Optomecha 7(3):1–15
16. Kelley CT (1999) Iterative methods for optimization. SIAM, Philadelphia, pp 135–137
17. Jiang BC, Shiau MY (1990) A systematic methodology for determining/optimizing a machine vision system's capability. Mach Vis Appl 3(3):169–182
18. Su TL, Chen HW, Hong GB, Ma CM (2010) Automatic inspection system for defects classification of stretch kintted fabrics. In: Proceedings of Wavelet Analysis and Pattern Recognition, pp 125–129
19. Ho SY, Huang HL (2001) Facial modeling from an uncalibrated face image using flexible generic parameterized facial models. IEEE.Trans Syst Man Cybern B 31(5):706–719
20. Chian TW, Kok SS, Lee SG, Wei OK (2008) Gesture based control of mobile robots. In: Proceedings of the Innovative Technologies in Intelligent Systems and Industrial Applications, pp 20–25
21. Li M, Milor L, Yu W (1997) Developement of optimum annular illumination: a lithography-TCAD approach. In: Proceedings of Advanced Semiconductor Manufacturing, pp 317–321
22. Lin CF, Wu CC, Yang PH, Kuo TY (2009) Application of Taguchi method in light-emitting diode backlight design for wide color gamut displays. J. Disp Tech 5(8):323–330
23. Yoo WS, Jin QQ, Chung YB (2007) A study on the optimization for the blasting process of glass by Taguchi method. J. Soc Korea Ind Sys. Eng 30(2):8–14
24. Wu Y, Wu A (2000) Taguchi methods for robust design. ASME Press, New York, pp 3–16

Chapter 6
Dense 3D Reconstruction in Multi-camera Systems

Soonmin Bae

Abstract The 3D information is useful in the security services and media production markets. However, it is difficult to achieve a reliable and dense 3D reconstruction from images. In this paper, we introduce a new approach to generating a dense 3D map from stereo images. We estimate the epipolar geometry and find matches using the geometric model. Given matches, we triangulate the 3D points and propagate the 3D information to neighboring pixels. For propagation, we use segmentation for a plane fitting and non-homogeneous optimization for an edge-preserving smoothing.

Keywords 3D reconstruction • Stereo

6.1 Introduction

Multi-camera systems become popular in recent years. Surveillance systems use them to monitor specific areas. When multi-camera systems need to track a moving target, it is essential to pass the 3D information to each other (Fig. 6.1).

Media production uses multi-camera techniques to broadcast sports. The growing 3D TV market will demand more advanced techniques.

However, it is difficult to achieve a reliable and dense 3D reconstruction from images. This paper introduces a new approach to generating a 3D map from non-rectified stereo images. We estimate the epipolar geometry find matches using the geometric model. Given matches, we triangulate the 3D points and propagate the 3D information to neighboring pixels. For propagation, we segment the input image and apply a plane fitting. In addition, we perform non-homogeneous optimization for an edge-preserving smoothing.

S. Bae (✉)
Intelligent Control Technology Team, Advanced Technology Center, Samsung Techwin, Seongnam, South Korea
e-mail: soonmin.bae@samsung.com; soonminb@gmail.com

R. Tutsch et al. (eds.), *Progress in Optomechatronic Technologies*,
Lecture Notes in Electrical Engineering 306, DOI: 10.1007/978-3-319-05711-8_6,

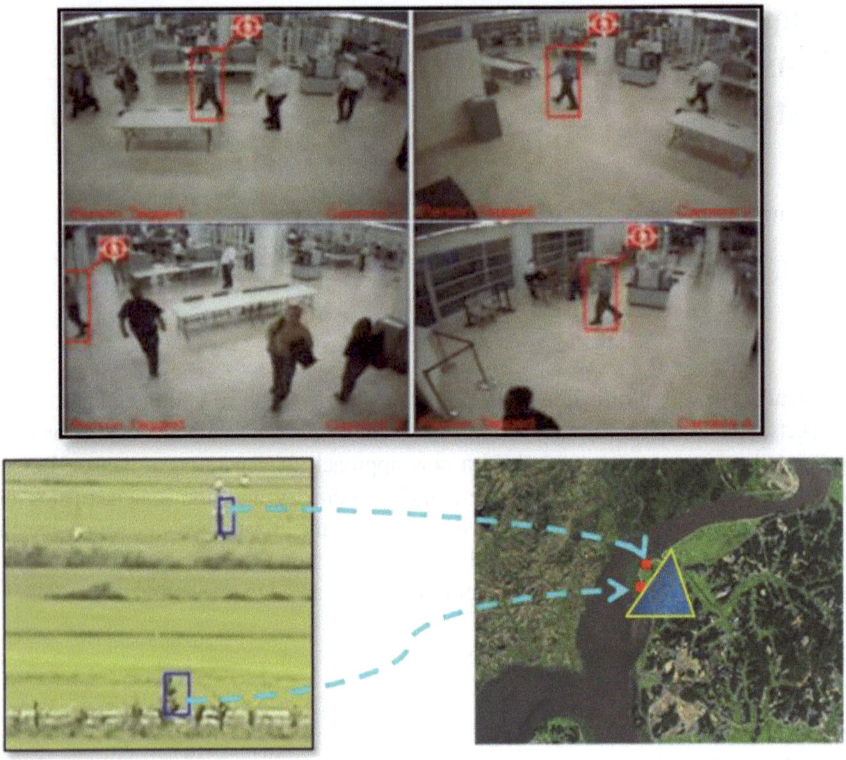

Fig. 6.1 Multi-camera surveillance systems

Although there exist previous methods which reconstruct 3D structures using stereo and segmentation [1, 4, 7], they deal with rather simple depth maps and rely on quite complicated algorithms.

6.2 Algorithm Description

For each pixel, we estimate a depth value. We call our depth estimation the depth map. We estimate the depth map in three steps. First, we estimate the depth value at corners. Then, we segment the image and apply a 3D plane-fitting to each segment given the sparse depth values. Finally, we propagate this depth measure to the rest of the image.

6.2.1 Epipolar Geometry Estimation and Triangulation

Given two non-rectified stereo images (Fig. 6.2), we estimate the epipolar geometry [3]. Since we assume the images are calibrated, we estimate the essential matrix. We lower the threshold for the corner detection to detect more matches (Fig. 6.3).

Fig. 6.2 Input images. They are not rectified, but calibrated

Fig. 6.3 Corner detection results (*red* and *green points*) and resulting matches (*blue lines*) satisfying the epipolar geometry

We triangulate the resulting matches and compute the initial 3D points. Although we increased the number of matches, they are points in the corners and do not cover the whole image. Therefore, we propagate the initial 3D points to the whole image (Fig. 6.4).

6.2.2 Segmentation and Plane-Fitting

To propagate the initial sparse depth value to the neighboring pixels, we segment the input image. After segmentation [2], we explicitly enforce the depth discontinuities in the segmentation boundaries and the depth continuities inside each segment using a plane-fitting. This results in sharp discontinuities and smooth planes (Figs. 6.5 and 6.6).

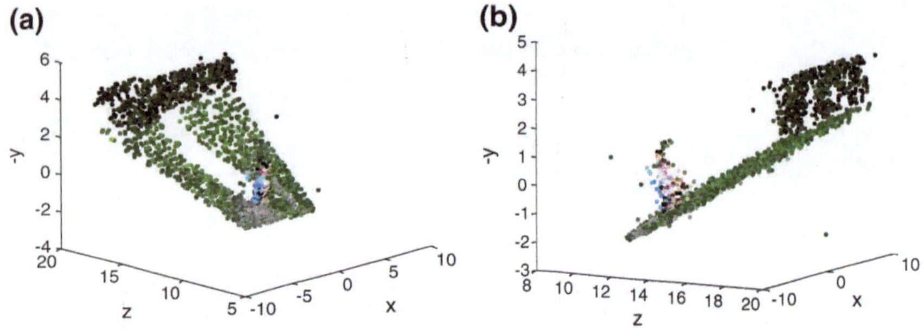

Fig. 6.4 The initial sparse 3D from two different views of the first example in Figs. 6.2 and 6.3. The horizontal rotation and vertical elevation in degrees are **a** −45 and 20, and **b** 60 and −10 respectively

Fig. 6.5 Segmentation results

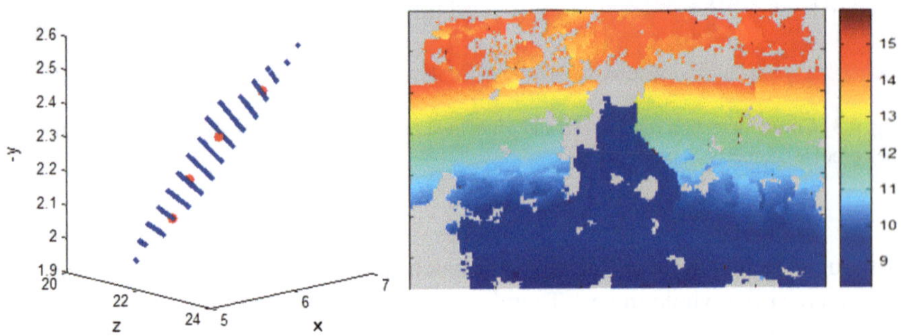

Fig. 6.6 Plane fitting example and plane fitting results. The *red values* means far and *blue* means near

To find a best-fitting plane, we use RANSAC (RANdom SAmple Consensus) [3] and PCA (Principal Component Analysis). We randomly select four initial depth values in a segment and estimate a corresponding plane using SVD (Singular Value Decomposition). We iterate the above process and find a best-fitting plane with a smallest fitting error. We register the plane if the plane satisfies more than 60 % of the initial depth values with less than 0.1 fitting error in the segment and if the area containing the initial depth values projected to the selected plane covers more than 20 % of the area of the segment projected to the plane.

However, to fit a plane, there should be at least three matches and there exist segments with less than three matches. To resolve this issue, we use an edge-preserving smoothing.

6.2.3 Edge-Preserving Smoothing

For the segments where a plane cannot be fitted, we propagate the initial 3D maps to the whole image using non-homogeneous optimization [5]. Our assumption is that the depth varies smoothly over the image except where the color is discontinuous. We propagate the depth value to the neighbors with similar color.

We impose the constraint that neighboring pixels p, q have similar depth if they have similar colors. We minimize the difference between the depth $D(p)$ and a weighted average of depth of neighboring pixels:

Non-homogeneous optimization:

$$E(D) = \sum (D(p) - \sum_{q \in N(p)} w_{pq} D(q))^2 + \sum \alpha_p (D(p) - ID(p))^2$$
$$\text{with: } w_{pq} \propto \exp\left(\frac{-(C(p) - C(q))^2}{\sigma_p^2}\right) \tag{6.1}$$

where $ID(p)$ is the initial depth measure from the plane fitting and σ_p is the standard deviation of the colors of neighboring pixels in a window around p. The window size used is 21×21. We have experimented both with setting the second term as hard constraints versus a quadratic data term, and have found that the latter is more robust to potential remaining errors in the depth measure.

6.3 Results

We have implemented our dense 3D reconstruction using Matlab. We crop the upper and lower 2 % of the depth map.

6.3.1 Depth Map Results

Figure 6.7 presents the depth map of the input images in Fig. 6.2. Figure 6.8 shows an additional example and result.

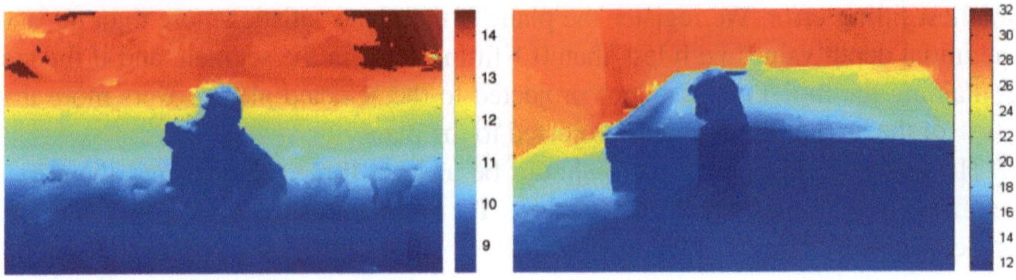

Fig. 6.7 The depth map results of the input images in Fig. 6.2. The *red* values means far and *blue* means near

Fig. 6.8 An additional result. The *red* values means far and *blue* means near

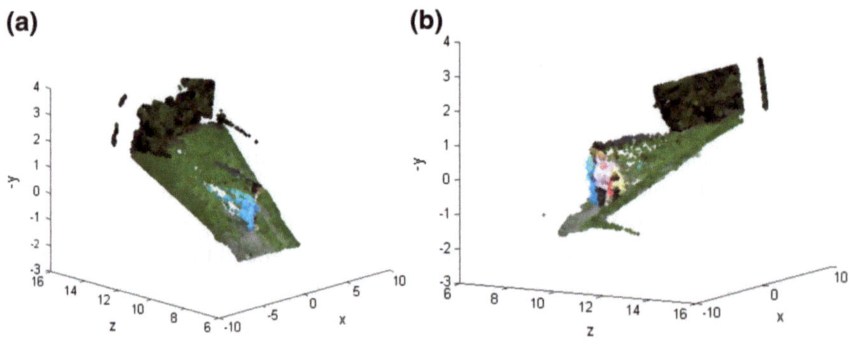

Fig. 6.9 The final dense 3D from two different views of the example in Fig. 6.4. The depth map clearly shows the depth discontinuities between the girl and bush and continuities in the grass and bush. The horizontal rotation and vertical elevation in degrees are **a** -45 and 20, and **b** 60 and -10 respectively

6.3.2 3D Results

Since we assume the camera is calibrated, we can compute the normalized coordinate for each pixel. As we multiply the depth value to the normalized coordinate, we achieve the 3D values for each pixel (Fig. 6.9).

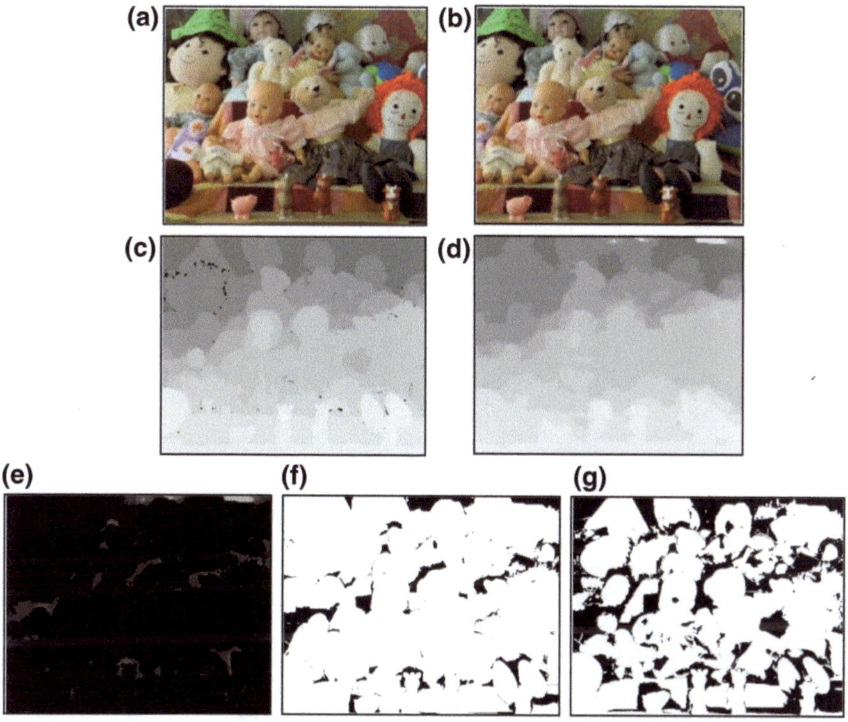

Fig. 6.10 Evaluation results. **a** Left image, **b** right image, **c** ground-truth disparities, **d** our result, **e** disparity differences, **f** pixels with less than 3 pixel disparity error, and **g** pixels with less than 1 pixel disparity error. 92 % pixels have less than a 3 pixel-error, and 85 % pixels have less than a 1 pixel-error. The median error ratio is 4.2 %

6.3.3 Evaluation

We evaluate the accuracy of our 3D using the Middlebury stereo datasets at http://vision.middlebury.edu/stereo/data/ [6]. Although the datasets are rectified, we do not rely on the fact that there is no rotation between the left and right cameras. The Middlebury stereo datasets provide the ground-truth disparities from a structured-lighting approach. We convert our depth map into disparity map using the following equation.

Conversion between depth and disparity:

$$Depth = \frac{focal\ length}{disparity} \tag{6.2}$$

Figures 6.10 and 6.11 show the results. They present that the corresponding disparity map from our 3D is very close to the ground-truth.

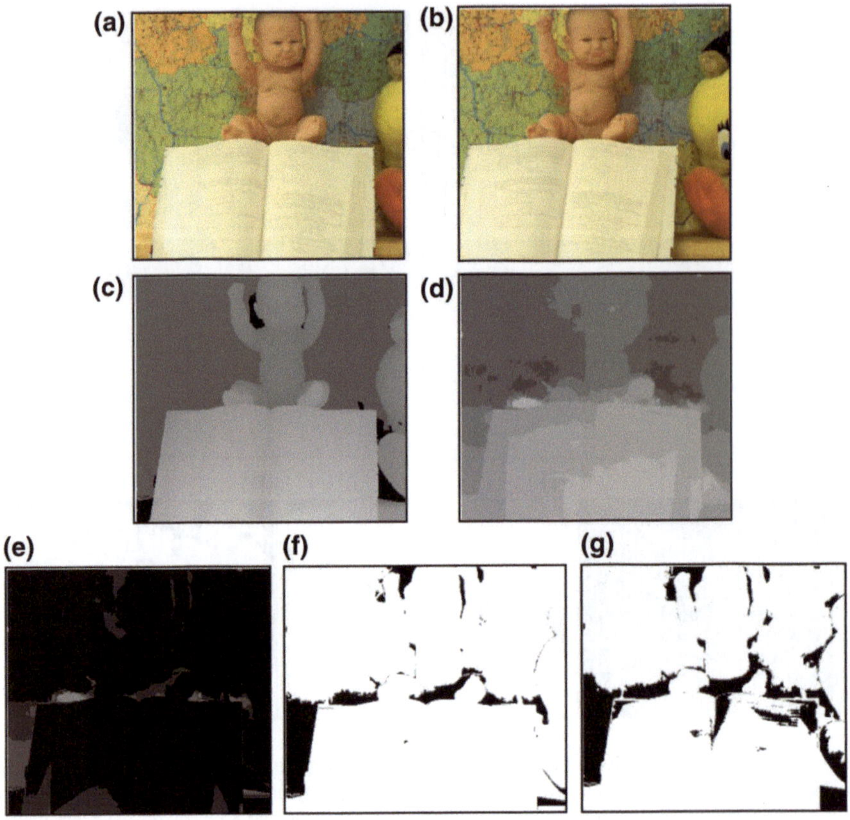

Fig. 6.11 More evaluation results. **a** Left image, **b** right image, **c** ground-truth disparities, **d** our result, **e** disparity differences, **f** pixels with less than 3 pixel disparity error, and **g** pixels with less than 1 pixel disparity error. 91 % pixels have less than a 3 pixel-error, and 72 % pixels have less than a 1 pixel-error. The median error ratio is 1.9 %

In Fig. 6.10, 92 % pixels have less than a 3 pixel-error, and 85 % pixels have less than a 1 pixel-error. The median error ratio is 4.2 %. In Fig. 6.11, 91 % pixels have less than a 3 pixel-error, and 72 % pixels have less than a 1 pixel-error. The median error ratio is 1.9 %.

6.4 Conclusion

This paper introduces a new approach to reconstructing a dense 3D from a sparse 3D. Given a pair of non-rectified images, we estimate sparse depth values at corners and propagate the 3D information to the neighboring pixels using a plane-fitting and edge-preserving smoothing. Our assumption is that depth is piece-wise smooth and particularly smooth where colors are similar. We verify the accuracy of our 3D using the ground-truth disparity maps.

We envision that the resulting dense 3D reconstruction will be of benefit to the security services and media production markets.

References

1. Bleyer M, Gelautz M (2005) Graph-based surface reconstruction from stereo pairs using image segmentation. Proc SPIE 5665:288–299
2. Comaniciu D, Meer P (2002) Mean shift: a robust approach toward feature space analysis. IEEE Trans Pattern Anal Mach Intell 24:603–619
3. Hartley R, Zisserman A (2000) Multiple view geometry in computer vision. Cambridge Press, Cambridge
4. Hong L, Chen G (2004) Segment-based stereo matching using graph cuts. In: IEEE Computer Society conference on computer vision and pattern recognition
5. Levin A et al (2004) Colorization using optimization. ACM Transactions on Graphics, In: Proceedings of ACM SIGGRAPH conference 23(3):689–694
6. Scharstein C, Pal C (2007) Learning conditional random fields for stereo. In: IEEE Computer Society conference on computer vision and pattern recognition
7. Strecha C et al (2003) Dense matching of multiple wide-baseline views. In: IEEE international conference on computer vision

Part II
Optical Metrology Including Measurement Systems of Microsphere, 3D Microscope and Large Surface

Chapter 7
Optimization of Coupling Condition in Distance Between the Sphere and the Tapered Fiber for Diameter Measurement of Microsphere by Using WGM Resonance

Atsushi Adachi, Masaki Michihata, Terutake Hayashi and Yasuhiro Takaya

Abstract In this study, it was proposed that whispering gallery mode (WGM) resonances was applied for diameter measurement of microsphere. WGM resonances were excited in a microsphere by introducing light through tapered fiber that thinned optical fiber to a few micrometers. Resonance wavelengths could be measured by monitoring intensity of the transmitted light of tapered fiber, since the light is strongly coupled to the microsphere at the resonance wavelengths. Therefore it is important to address the coupling conditions in order to measure WGM resonances accurately. Thus, optimization of the coupling is key issue to determine the diameter of a microsphere. Numerical analysis was implemented to estimate the coupling coefficients of each modes of WGM. Then the coupling was conducted experimentally as parameter was distance between the sphere and the tapered fiber.

Keywords WGM resonance • Diameter measurement • Tapered fiber • CMM probe

7.1 Introduction

Thanks to progress processing technologies, many processing technologies have achieved an accuracy of nanometer order and thus met demands for miniaturization of products. However measurement technologies to evaluate three-dimensional form of the microparts less than several millimeters in size are still in development. Conventionally, coordinate measuring machines (CMMs) are used in production process as the standard instrument for three-dimensional absolute form. CMM acquires spatial coordinates of the surface of measured objects by moving a probe

A. Adachi (✉) · M. Michihata · T. Hayashi · Y. Takaya
Department of Mechanical Engineering, Osaka University, Osaka, Japan
e-mail: adachi@optim.mech.eng.osaka-u.ac.jp

R. Tutsch et al. (eds.), *Progress in Optomechatronic Technologies*,
Lecture Notes in Electrical Engineering 306, DOI: 10.1007/978-3-319-05711-8_7,

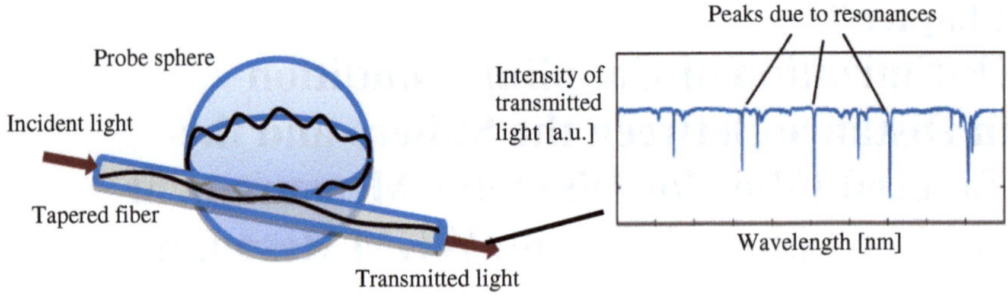

Fig. 7.1 WGM resonances exited by tapered fiber [2]

system. The probe system consists of a probe shaft and a probe stylus (generally sphere). To measure the dimension of the microparts it is need to miniaturize the probe stylus as well as whole CMM systems. Diameter of probe stylus should be less than 100 μm, which is necessary to be evaluated strictly because the CMMs acquire surface coordinate of measured object using probe radial offset [1]. Then the measurement accuracy for the probe diameter must be less than 10 nm.

The goal of this study is the establishment of the new measuring principle to evaluate the diameter of a microsphere with nanometer order accuracy. The proposed measurement method of diameter of a microsphere uses travelling light in the sphere. After the light enter a sphere, the sphere serves as an optical cavity, so whispering gallery mode resonance (WGM resonance) is excited, which is the optical mode going around the surface layer in the sphere for specific wavelength. The wavelength of WGM resonance is used to measure the diameter. Since there are various modes in WGM resonances the coupling strength at each mode is calculated to identify the each mode inside the sphere into the found each resonance wavelength. In this paper the diameter measurement principle is studied, i.e. the technique to determine appropriate resonance wavelengths is investigated.

7.2 Diameter Measurement Using WGM Resonance

7.2.1 Excitation of WGM Resonance with Tapered Fiber

Light entered inside a dielectric microsphere repeats total internal reflection at surface of the sphere. The light is confined and resonated in the microsphere if circumference of the microsphere is equal to the integral multiple of wavelength of the travelling light. This phenomenon is so-called WGM resonance (Fig. 7.1). WGM resonance is sensitive to morphological change such as size, shape and refractive index of microsphere. For optical coupling to microsphere, tapered fiber is used. At tapered section, the total internal reflection at boundary plane between the fiber and surrounding air yields the evanescent light on surface of the tapered fiber. Light is coupled to a microsphere via evanescent light. When the wavelength of incident light to the tapered fiber is swept, the intensity of light transmitted through the tapered fiber decreases where strong coupling to WGM resonance is occurred at WGM resonance wavelength. Thus the dips (resonance peaks) are appeared in the graph of transmitted light by this coupling (Fig. 7.1). The wavelengths at this resonance peaks are WGM resonance wavelength in a microsphere.

7.2.2 Measurement Principle of Microsphere Diameter Using WGM Resonance

There are two kinds of whispering gallery mode—transverse magnetic (TM) and transverse electric (TE). In the case of TM mode, the magnetic field is tangent to the surface of the sphere and perpendicular to the direction of propagation. The electric field has components normal to the surface of sphere and parallel to the direction of propagation. In the case of TE mode, these properties are reversed. The wavelengths of these WGM resonances are given by Eq. (7.1) [3].

$$\lambda_{qlm} = 2\pi a_s n_s \left[l + \frac{1}{2} + \alpha_q \sqrt[3]{\left(l + \frac{1}{2}\right)/2} - \Delta \pm \varepsilon^2 \frac{l - |m|}{2} \right]^{-1} \tag{7.1}$$

where a_s is the sphere's radius, n_s is the refractive index of sphere, n_0 is the refractive index of surrounding medium, l is the polar mode number (integer with $l \geq 1$), q is the radial mode number (integer with $q \geq 1$), m is the azimuthal mode number (integer with $|m| \leq l$), α_q is the absolute value of the qth zero of the Airy function, ε is the eccentricity of sphere and Δ gives polarization shift of the WGM resonance wavelength: $\Delta_{TM} = \gamma^{-1}\left(\gamma^2 - 1\right)^{-\frac{1}{2}}$, and $\Delta_{TE} = \gamma\left(\gamma^2 - 1\right)^{-\frac{1}{2}}$, where $\gamma = n_s/n_0$. The radial mode number illustrates how interior light travels. The light of $q = 1$ travels at the nearest position to the surface of the sphere. The polar mode number illustrates wave number contained in the circumference. The azimuthal mode number illustrates how broad light spread. If absolute value of the azimuthal mode number is maximum, the broadening of the light is minimum. The radial components of the electric field in TM mode WGM for microsphere in diameter of 20 μm was calculated. Figure 7.2 shows cross-sectional electromagnetic field in a sphere. The light travelling interior relative to the surface of a sphere and the Eq. (7.1) shows the wavelength of the light penetrating interior of a sphere and travelling in a sphere. In a perfect sphere, the azimuthal modes are frequency degenerate. Therefore the diameter of sphere can be determined from Eq. (7.1) based on the measured WGM wavelengths and the radial mode number q. However in the case of distorted sphere, the degeneracy is broken and the resonance wavelengths are divided according to the azimuthal mode numbers. At that condition, there are many resonance peaks appeared. Therefore the determination of the radial mode number q and increased resonance wavelengths by different azimuthal mode number m are the problem of the measurement.

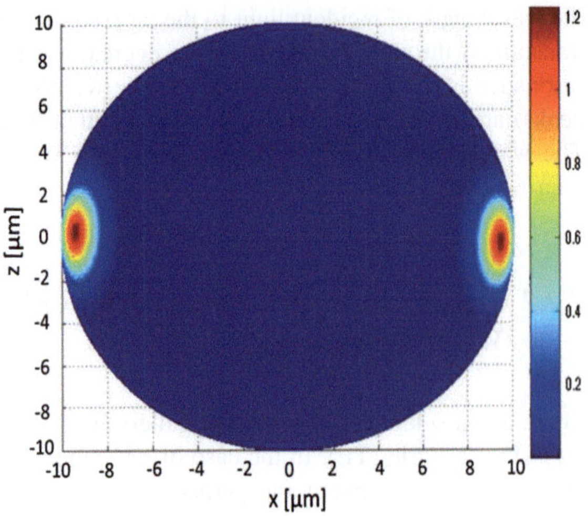

Fig. 7.2 Radial components of the electric field of TM mode WGM resonance (x-z plane)

7.3 Numerical Analysis of Coupling Coefficient

The depth of the resonance peak appearing in the transmitted light intensity linearly-increase with the coupling coefficient κ given by Eq. (7.2) [3]. The coupling coefficient is calculated to investigate the depth of resonance peak at each mode.

$$\kappa = \frac{\omega \varepsilon_0}{4P}\left(n_s^2 - n_0^2\right) \iiint_{V_s} \left(E_f \cdot E_s^*\right) \cos\left(m\phi - r\beta_f \sin\phi\right) r^2 dr d\theta d\phi \quad (7.2)$$

V_s is the volume of a sphere, E_f is the electric field of a tapered fiber, E_s^* is the complex transposition of the electric field of a sphere and β_f is the propagation constant of the light in a tapered fiber. The calculation procedure of the coupling coefficient has three steps. First, the resonance wavelength of each mode is calculated from Eq. (7.1). Second, the electric field of a sphere and a tapered fiber are calculated using resonance wavelength acquired in first step. Third, the coupling coefficient of each mode is calculated by Eq. (7.2). Conditions of calculation are the following. The microsphere's diameter was 20 μm and refractive index was 1.501. Diameter of the tapered fiber was 1 μm. The microsphere was contacted in a tapered fiber. The calculation result of the coupling coefficients at different azimuthal mode number (TM mode WGM resonance, radial mode number $q = 1$ and polar mode number $l = 54$) is shown in Fig. 7.3. The coupling coefficient at $l = m = 54$ which the travelling light does not broaden is the maximum of all azimuthal modes. The light travels in the tapered fiber couples most strongly to this mode. The same results are obtained at other polar mode number l. Then it was investigated the dependence on the radial mode number of this $m = l$ mode for changes of coupling coefficients. The calculation result of coupling coefficients at different radial mode number (TM mode WGM resonance and $m = l$) existing in

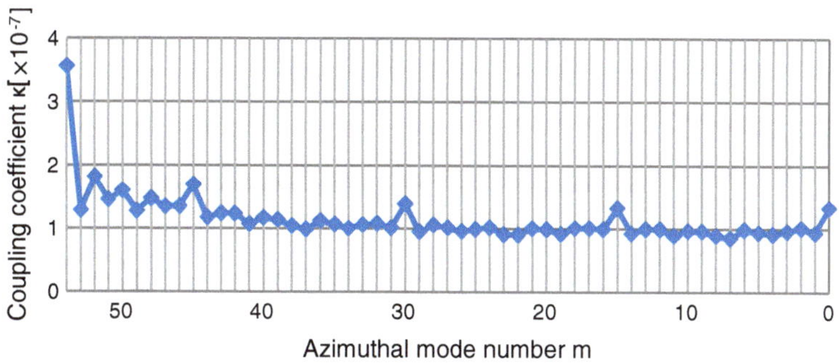

Fig. 7.3 TM mode WGM resonance coupling coefficient at different azimuthal mode number(q = 1, l = 54)

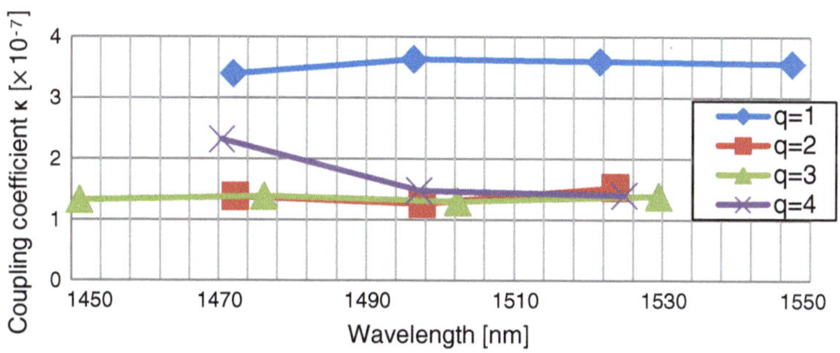

Fig. 7.4 TM mode WGM resonance coupling coefficient at different radial mode number

1450–1550 nm wavelength range is shown in Fig. 7.4 which insists that the light couples most strongly to the mode with radial mode number $q = 1$ where the light travels close to the surface of sphere of all radial modes.

As a results, the resonance peaks at $q = 1$ and $m = l$ mode is deeper than other modes. Same calculation results are obtained for TE mode WGM. From this the resonance peaks caused by TM mode or TE mode or both TM mode and TE mode WGM resonance would appear depending on the polarization of incident light into a sphere. Therefore Eq. (7.1) simplified to Eq. (7.3) because the resonance peaks at $q = 1$ and $m = l$ is significantly appearing.

$$\lambda_{qlm} = 2\pi a_s n_s \left[l + \frac{1}{2} + \alpha_1 \sqrt[3]{\left(l + \frac{1}{2} \right) / 2} - \Delta \right]^{-1} \tag{7.3}$$

The simultaneous equation is written by choosing the wavelength of the adjacent resonance peak appearing greatly in the Eq. (7.3). The diameter of microsphere is calculated from this simultaneous equation.

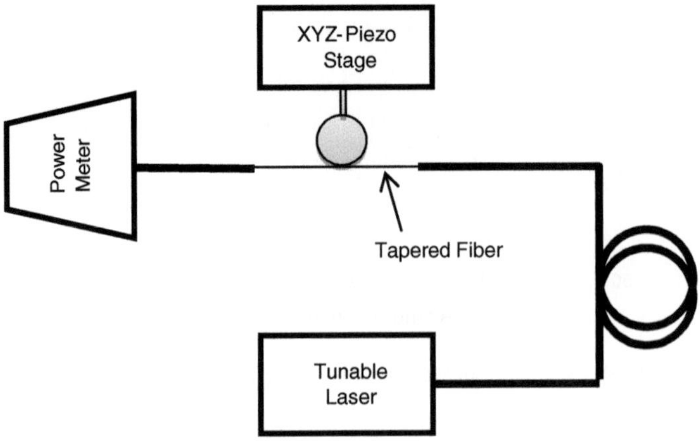

Fig. 7.5 Schematic of the experiment system

7.4 Diameter Measurement of Glass Micro-Probe

7.4.1 Experimental Setup

Figure 7.5 shows the experimental setup. Material of microsphere is optical glass bk7 (refractive index of 1.501). The microspheres are made by using surface tension. The heating the optical cylinder glass made itself sphere by surface tension. The sphere is mounted on a XYZ-piezo-stage to precisely control distance between the tapered fiber and microsphere. The microsphere and tapered fiber are observed by CCD camera (not shown in Fig. 7.5). The wavelength-tunable laser was connected to the one side of the tapered fiber. The power meter was connected other side of the tapered fiber in order to measure the intensity of transmitted light. Diameter of tapered fiber was 1 μm. Diameter of microspheres are 90 and 130 μm, respectively, which was roughly estimated by optical microscope image.

7.4.2 Experimental Results

Intensity of light transmitted through the tapered fiber is measured to obtain the WGM resonance peaks. After the microsphere was contacted in the tapered fiber, the wavelength of the laser was swept to find the WGM resonance peaks. The measured intensities were normalized by intensities without microsphere. Figure 7.6 shows the measured intensity of transmitted light for a microsphere of diameter 90 and 130 μm, respectively. As shown, WGM resonance peaks were observed, however, it is difficult to identify mode number against the resonance peaks because different modes appear as the resonance peaks.

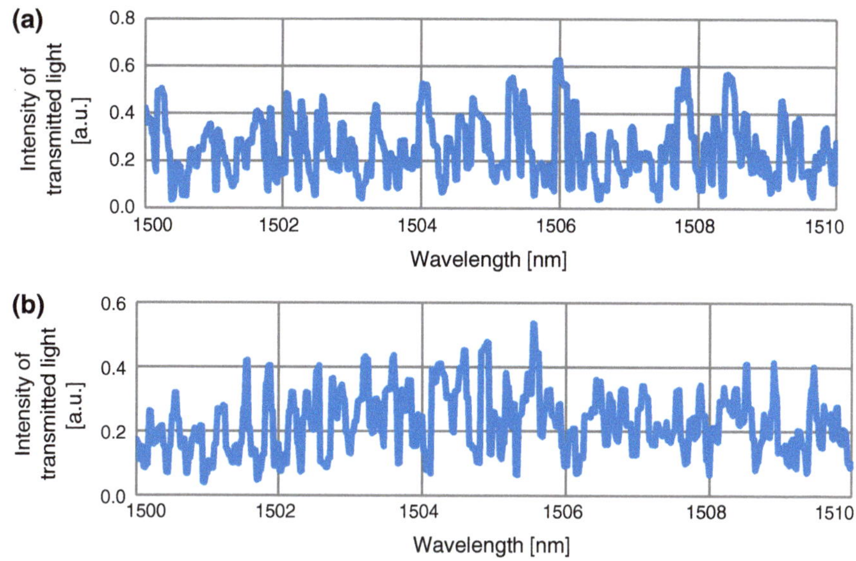

Fig. 7.6 Intensity of transmitted light in the case of microsphere was contacting in tapered fiber. **a** Sphere of $\phi 90\,\mu\text{m}$ **b** Sphere of $\phi 130\,\mu\text{m}$

Equation (7.2) implies that coupling coefficient was superimpose of electromagnetic fields derived both the tapered fiber and the sphere. In this point of view, the distance between the microsphere and the tapered fiber may affect coupling strength, because evanescent light field was exponentially decayed from surface of the tapered fiber. Therefore, it was thought that the modes of weak coupling coefficient would not appear as the resonance peak in the case that there is a gap between the microsphere and the tapered fiber. Then, the distance between the microsphere and the tapered fiber was adjusted by the piezo-stage. Figure 7.7 shows the resonance peaks after the adjustment of the distances, where the number of the resonance peaks was clearly reduced successfully.

Two microspheres which differ in size were measured in this experiment. The resonance peaks of the larger microsphere were shallower than that of the smaller microsphere. From calculation of electric field distribution, intensity maximum is slightly inner than surface of sphere. The distance of the intensity maxima and surface of the sphere becomes large with increasing diameter of microsphere. Therefore the larger microsphere needed to be approached more closely to the tapered fiber for the observation of resonance peaks. It means the optimized distance for measurement of resonance peaks is different depending on the diameter of microsphere.

Finally the diameter was estimated from the measured wavelength of WGM resonance. Table 7.1 shows the diameter calculated by the wavelengths of the resonance peaks as marked with circles in Fig. 7.7. These resonance peaks were assumed to be caused by TE mode WGM resonance, because polarization of light in the microsphere was not known. In this measurement resonance peaks by other modes were still appeared, so theoretical value of wavelength interval for

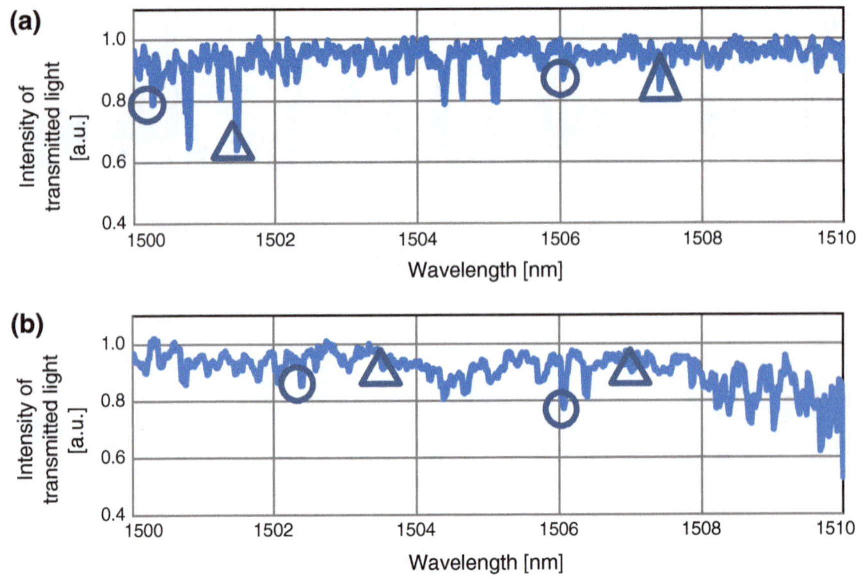

Fig. 7.7 Intensity of transmitted light in the case of sphere was not contacting in tapered fiber. **a** Sphere of $\phi 90\,\mu\text{m}$ **b** Sphere of $\phi 130\,\mu\text{m}$

Table 7.1 Resonance wavelength and calculated diameter

	Resonance wavelength (nm)		
	Shorter one	Longer one	Calculated diameter (μm)
Sphere of $\phi 90\,\mu\text{m}$	1500.26	1505.72	89.041
Sphere of $\phi 130\,\mu\text{m}$	1502.38	1506.04	132.586

microsphere of 90 and 130 μm in diameter. Theoretically, the resonance peaks were supposed to appear at interval of about 5.4 nm for diameter of 90 μm and about 3.7 nm for diameter of 130 μm. On the basis of this analysis, the wavelengths of WGM resonance were chosen to calculate a diameter. As a result, 89.041 μm for the sphere of 90 μm in diameter and 132.586 μm for the sphere of 130 μm in diameter were obtained. From this, resonance peaks of strong coupling coefficient were expected to exist when there was a gap between the microsphere and the tapered fiber. Therefore, the capability of diameter measurement is shown.

7.5 Discussion

Modes of WGM resonance was classified into two modes depending on the polarization: TE and TM mode, which must be considered beyond the number of modes. In this measurement, it is difficult to determine which types of WGM resonance is

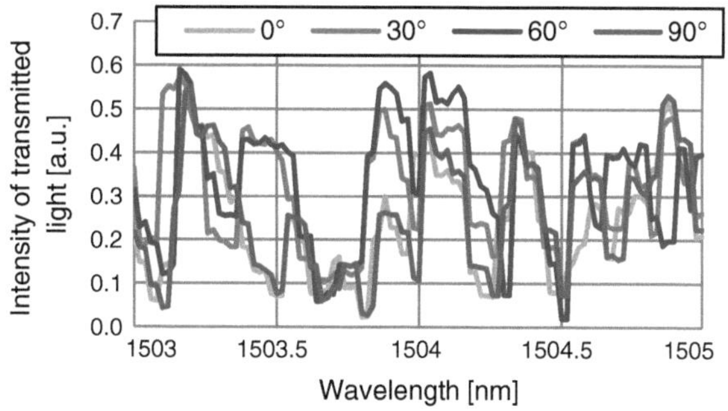

Fig. 7.8 Intensity of transmitted light at different incident polarization angle

excited. These modes of WGM resonance are determined by polarization of incident light into the sphere. The diameter of the tapered fiber used in this experiment was 1 μm, therefore single mode propagation in the tapered fiber at wavelength range over 1,500 nm is considered. Electric field in the single mode fiber is axial symmetry with independent on incident polarization direction. Therefore, the excited WGM resonance may not depend on the incident condition by the tapered fiber. To check this phenomenon, the polarization rotator was installed before the laser is coupled into the tapered fiber to control the polarization direction. The intensity of the transmitted light was measured at different angles of polarization direction (Fig. 7.8). The angle of the polarization rotator was changed in steps of 30°. There are a little intensity variation caused by the noise, but the wavelengths of resonance peaks did not depend on the change of the incident polarization. The results imply that both TM and TE mode of WGM resonance are excited simultaneously. Resonance peaks of TM mode generally appear at longer wavelength than one of TE mode. Then the wavelength of TM mode WGM resonance is calculated from the diameter obtained in the measurement. Examples are about 1.5 nm longer for diameter 90 μm and 0.95 nm longer for diameter of 130 μm respectively. As seen the triangle marks in Fig. 7.7 there are the candidates of resonance peaks of TM mode. Since the peak position was in good agreement with theoretical estimation, it is considered that WGM resonance of both the TM and TE mode were excited in this experiment. Therefore the resonance peaks of strong coupling coefficient caused by both TM mode and TE mode WGM resonance appear if the distance between the sphere and the tapered fiber is optimized.

7.6 Conclusion

The new measurement method of diameter of a microsphere is proposed. The numerical analysis, experiment of excitation of WGM resonance and applying proposed measurement method are carried out. In the numerical analysis, the specific microsphere modes couple strongly, in which the light travels close to the surface of sphere ($q = 1$) without broadening ($|m| = l$). Therefore the resonance peaks of these modes appear. Two microsphere which differ in size are measured in experiment. In the experiment when the microsphere was contacting in tapered fiber, there were many resonance peaks appearing and it was difficult to identify the resonance peaks of such mode. In the experiment when the microsphere was not contacting tapered fiber, the number of resonance peaks appearing were reduced. Exsisting the resonance peaks at intervals of wavelength distance calculated from spherical diameter, the capability of diameter measurement was shown. However, the resonance peaks by the other modes are still appeared. Therefore the mode number has to be identified by using different parameters from dip depth of resonance peaks as a future task.

Acknowledgements This work was financially supported by JSPS Grant-in-aid for challenging exploratory research (No. 24656102) and also Mitutoyo association for science and technology (MAST).

References

1. Kung A et al (2007) Ultraprecision micro-CMM using a low force 3D touch probe. Meas Sci Technol 18:319–327
2. Shainline J et al (2010) Broad tuning of whispering gallery modes in silicon microdisks. Opt Express 18:14345–14352
3. Humphrey MJ (2004) Ph.D thesis, University of Maryland

Chapter 8
UV Shadow Moiré for Specular and Large Surface Shape Measurement

Changwoon Han

Abstract Three dimensional shape measurement techniques using optical methods are extensively applied in many fields of industry for product development and quality control. High resolution, fast measurement, large vertical and horizontal detecting area, and being applicable regardless of materials are valuable factors for the techniques. Recently there are strong demands to develop the technique which can measure specular and large surface quickly in production process for quality control purpose. Shadow moiré can detect a large area due to the using a non-coherent light and measure in shorter time than scanning method but cannot be applied to specular surface. In the study, optical 3-D shape measurement technique based on shadow moiré is proposed. By adapting UV light instead of visible white light in shadow moiré, the new technique can measure even specular surface which was the limit of the existing shadow moiré. Also it can take full advantage of the heritage of shadow moiré; detecting a large area due to the using a non-coherent light and measuring in shorter time than scanning method. It can also increase vertical measuring range.

Keywords 3-D shape measurement • Shadow moiré • UV light • Specular surface

8.1 Introduction

Three dimensional shape measurement techniques using optical methods are extensively applied in many fields of industry for product development and quality control. High resolution, fast measurement, large vertical and horizontal detecting area, and being applicable regardless of materials are valuable factors for the techniques.

C. Han (✉)
Korea Electronics Technology Institute, Seongnam, South Korea
e-mail: cw_han@keti.re.kr

R. Tutsch et al. (eds.), *Progress in Optomechatronic Technologies*,
Lecture Notes in Electrical Engineering 306, DOI: 10.1007/978-3-319-05711-8_8,

Table 8.1 Comparsion of shape measurment techniques

	Detecting area			Measurement	
Technique	Horizontal	Vertical	Resolution	Type	Speed
OT [2]	<550 mm	<1 mm	~20 μm	Scanning	60 cm^2/s
PSI [3]	<14.5 mm	~138 nm	~0.1 nm	Full-field	<1 s
WSI [3]	<14 mm	<20 mm	~0.1 nm	Scanning	<135 μm/s
CSM [4]	<1.6 mm	<10 mm	~20 nm	Scanning	Low
ESPI [5]	<50 mm	Variable	0.01~0.1 μm	Full-field	<3.5 s
DIC [5]	<70 mm	Variable	10^{-5} × detecting area	Full-field	depending on frame-rate
SM [6]	<600 mm	<10 mm	~2.5 μm	Full-field	<1 s

Laser scanning, moiré method, laser speckle pattern sectioning, interferometry, photogrammetry, and laser tracking system are some examples for the technique [1]. There isn't the technique able to cover all the measuring factors for all the application. Optimal technique should be selected driven by the purpose of application.

Traditional applications of shape measurement technique in electronic industry are the automated optical inspection (AOI) on the surface of printed circuit boards, which is an optically diffusive, or mat, surface. But recently there are strong demands to develop the technique which can measure specular and large surface quickly in production process for quality control purpose. Large LCD/OLED display, photovoltaic panel, and flexible electronics from roll-to-roll process are some example for the application.

In this study, a new optical 3-D shape measurement technique based on shadow moiré is introduced. By adapting UV light instead of visible white light in shadow moiré, it was shown that it is possible to measure a specular and large surface quickly and also to increase the vertical detecting area.

8.2 3-D Shape Measurement Technologies Comparison

Optical triangulation (OT), phase shifting interferometry (PSI), white-light scanning interferometry (WSI), confocal scanning microscopy (CSM), electronic speckle pattern interferometry (ESPI), digital image correlation (DIC), or shadow moiré (SM) are typical techniques used in electronic industry for shape measurement. To compare the techniques each other, horizontal and vertical detecting area, resolution, and measurement speed are investigated from commercial products which are intended for electronic application and summarized at Table 8.1.

Maximum horizontal detecting areas change widely from 14 to 600 mm for WSI and SM, respectively. Measurement resolution also changes from 0.1 nm to 20 μm; PSI, WSI, and CSM have high resolutions while OT, DIC, and SM have low resolutions. Measurement speeds are depending on measurement type; scanning type measurements such as OT, WSI, and CSM inherently have low speed.

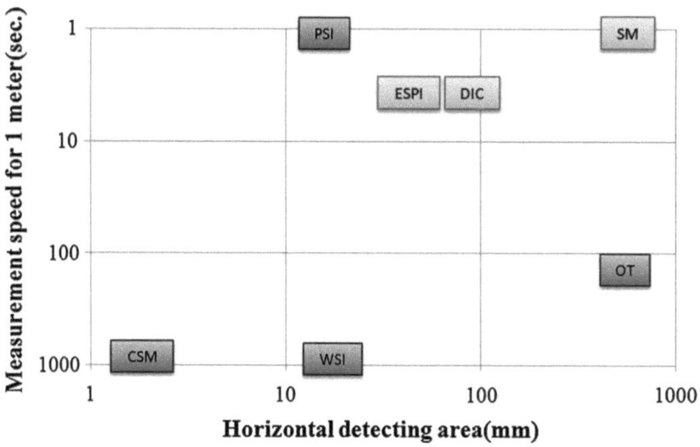

Fig. 8.1 Measurement speed versus horizontal detecting area for shape measurement techniques

Horizontal detecting areas and measurement speeds are graphically compared at Fig. 8.1. From the figure, it could be said that if large horizontal detecting area and high speed are required, ESPI, DIC, or SM could be a potential solution.

But one of disadvantages of ESPI, DIC, and SM is that they are only applicable to diffusive surfaces, not to specular surfaces. If one want to measure the shape of metal (Cu, Al) or polymer (PI, PEN), ESPI, DIC, or SM cannot be utilized.

In the next section, a novel shape measurement technique are proposed, which can be applied to both specular and diffusive surfaces even though having large horizontal detecting area and high measurement speed.

8.3 Shadow Moiré with UV Light

8.3.1 Principle of Shadow Moiré

Figure 8.2 illustrates a shadow moiré system. A linear reference grating of pitch g is positioned adjacent to the specimen surface. A light source illuminates the grating and the specimen at angle α, and a camera receives light at angle β, which is scattered in its direction by the diffusive specimen surface.

The governing equation is

$$z(x,y) = \frac{g}{\tan\alpha + \tan\beta} N(x,y) \tag{8.1}$$

where $z(x, y)$ is the distance to the surface of the specimen at each x, y point, and $N(x, y)$ is the moiré fringe order at the same point [7].

The contrast of shadow moiré fringes is actually controlled by the combined effects of the *Talbot* distance and the aperture of an imaging system [8]. As z increases, the contrast varies cyclically (the Talbot effect) with a gradually

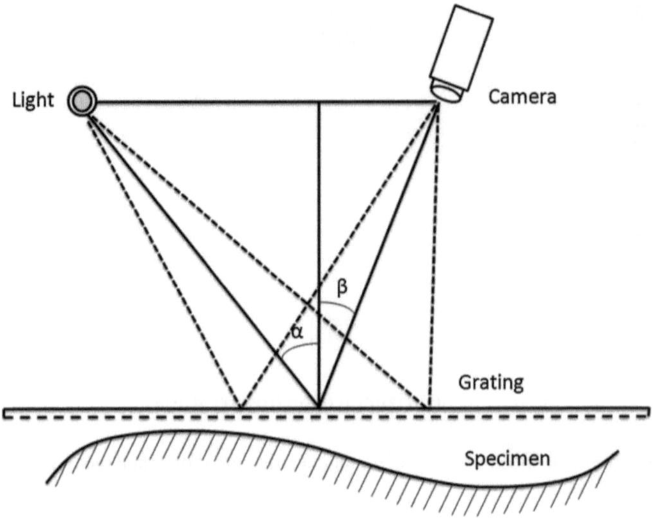

Fig. 8.2 Conventional configuration of shadow moiré system

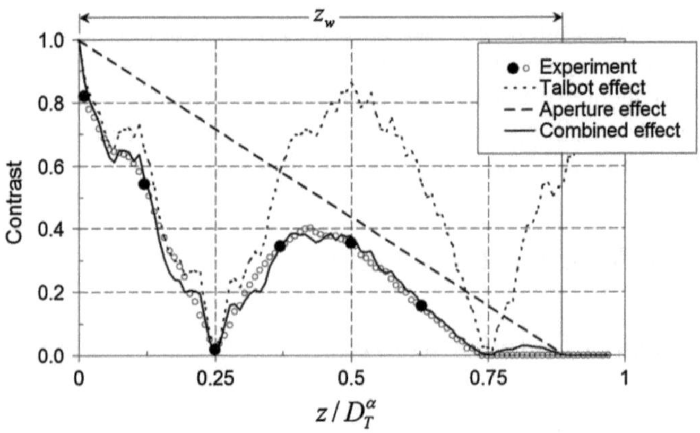

Fig. 8.3 Example of fringe contrast of shadow moiré showing the effect of Talbot and aperture effect [8]

decreasing amplitude (the aperture effect) as shown in Fig. 8.3. The Talbot effect is controlled by the diffraction of light at the grating and the Talbot distance for oblique illumination, D_T^α, is [8]

$$D_T^\alpha = \frac{2g^2}{\lambda} \cos^3 \alpha. \tag{8.2}$$

These effects are critically important for high-sensitivity measurements, where fine gratings and large angles of incidence are employed.

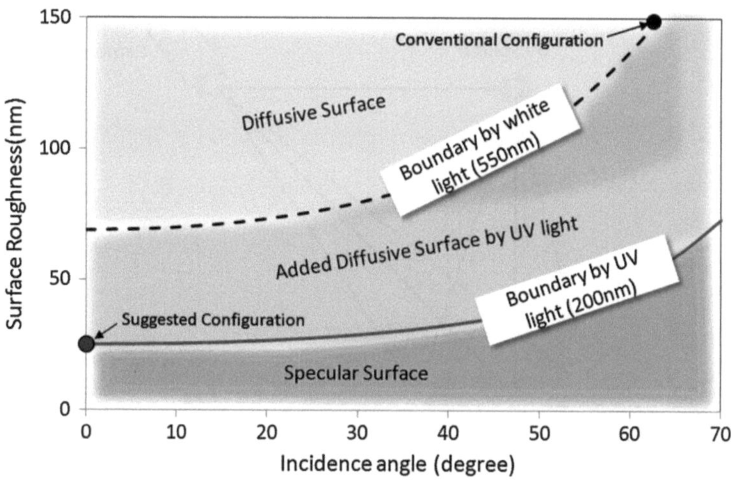

Fig. 8.4 Illustration for advantages of using UV light and normal illumination

8.3.2 Shadow Moiré with Ultraviolet Light

Rayleigh roughness criterion defines the condition of specular surface as [9]

$$h < \frac{\lambda}{8\cos\theta} \tag{8.3}$$

where h is the roughness of surface, λ is the wave length, and θ is the angle of incidence light. From Eq. (8.3), roughness of surface is relatively defined by the wavelength and incidence angle of light. The shorter the wavelength and the smaller the incidence angle of light, the criterion for specular surface is decreasing.

When we have a shadow moiré with visible white light ($\lambda_{average} = 550$ nm) and 63° illumination ($\theta = 63°$), Rayleigh criteria becomes 150 nm. In other words, shadow moiré cannot measure the surface less than 150 nm roughness. But if we change the light into ultraviolet ($\lambda = 200$ nm) and normal illumination ($\theta = 0°$), Rayleigh criteria becomes 25 nm. The roughness requirement becomes much less reduced. The advantage using UV light and normal illumination is illustrated in Fig. 8.4.

The suggested UV-based shadow moiré as shown in Fig. 8.5 can measure even specular surface which was the limit of the existing shadow moiré technique. Also it can take full advantage of the heritage of shadow moiré; detecting a large area due to the using a non-coherent light and measuring in shorter time than scanning method. As sensitivity of shadow moiré does not depend on the wavelength of light, measuring sensitivity and phase shifting technique can be utilized identically.

One more benefit of using UV light is increasing vertical measuring range. The vertical measuring range is governed by Talbot distance (Eq. 8.2) of shadow moiré system. In conventional configuration [7] of shadow moiré with visible white light ($\lambda_{average} = 550$ nm), 63° illumination ($\theta = 63°$), and 0.1 mm grating pitch

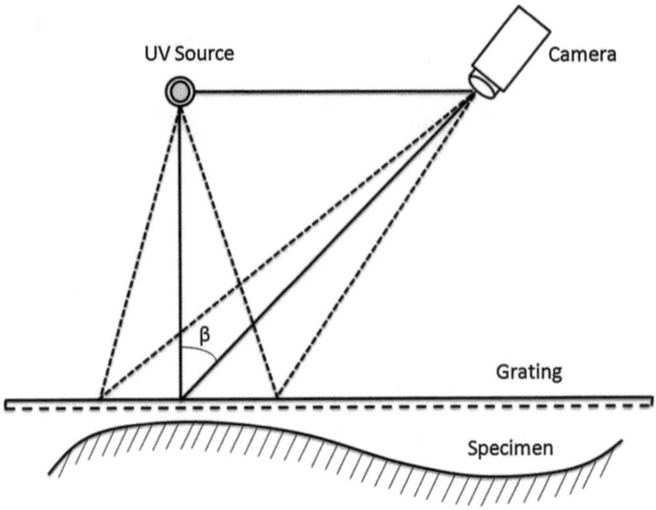

Fig. 8.5 Suggested configuration of UV-based shadow moiré system

(g = 0.1 mm), Talbot distance becomes 3.25 mm. But if we change the configuration into ultraviolet light ($\lambda = 200$ nm), normal illumination ($\theta = 0°$), and the same grating pitch ($g = 0.1$ mm), Talbot distance increase into 100 mm, which is more than thirty times larger than the conventional configuration's.

8.4 Conclusions

Recently there are strong demands to develop the technique which can measure specular and large surface quickly in production process for quality control purpose. Shadow moiré can detect a large area due to the using a non-coherent light and measure in shorter time than scanning method but cannot be applied to specular surface.

A novel optical 3-D shape measurement technique based on shadow moiré was introduced. By adapting UV light instead of visible white light in shadow moiré, the new technique can measure even specular surface which was the limit of the existing shadow moiré. Also it can take full advantage of the heritage of shadow moiré; detecting a large area due to the using a non-coherent light and measuring in shorter time than scanning method. One more benefit of using UV light is increasing vertical measuring range.

References

1. Chen F, Brown GM, Song M (2000) Overview of three dimensional shape measurement using optical methods. Opt Eng 39:10–22
2. Parmi. www.parmi.com, HS60L
3. Zygo. http://www.zygo.com, Newview 700p for PSI, NewView 7300 for WSI
4. Luniak M et al (2006) Optical characterization of electronic packages with confocal microscopy. Elec Sys Tech Conf, 1318–1322
5. Dantec Dyanmics. http://www.dantecdynamcis.com, Q-300TCT for ESPI, Q-400TCT for DIC
6. Akrometrix. http://www.akrometrix.com, PS600
7. Han C, Han B (2006) High sensitivity shadow moiré using nonzero-order Talbot distance. Exp Mech 46:543–554
8. Han C, Han B (2005) Contrast of shadow moiré at high order Talbot distances. Opt Eng 44(2005)
9. Rayleigh L (1945) The theory of sound. Dover, New York

Chapter 9
Development of Measurement System for 3D Microscope

Yeon-Beom Choi, Kee-Jun Nam, Woo-Jun Kim and Doo-Hee Jung

Abstract In this paper, we propose a measurement system for three-dimensional microscope which is designed to move the objective lens in two directions polar and azimuth angle. By only two measurements of adjusting polar angle with the controller and measuring the length of the object through camera, we can get three-dimensional information of the object like height, step and angle. Servo-controllers, PC interface, image acquisition software and image processing software have been developed for this system and experimental results show the effectiveness of the proposed system.

Keywords 3D measurement • Microscope • Polar angle • Azimuth angle

9.1 Introduction

The optical microscope, since its first appearance, has come to a great level of development with repeated evolution and innovation, for which at the present still new theory and technology are actively discussed. Likewise in the industry, its range of use has continued to expand, and its market has been steadily increasing as well.

Y.-B. Choi · K.-J. Nam · D.-H. Jung (✉)
Department of Electronics Engineering, Korea Polytechnic University, Gyeonggi-do 429-792, South Korea
e-mail: doohee@kpu.ac.kr

Y.-B. Choi
e-mail: choiyb0225@nate.com

K.-J. Nam
e-mail: kjnam7@naver.com

W.-J. Kim
Glonik, 806 Partners Tower 1 Cha, Gasan-dong, Geumcheon-gu, Seoul 153-802, South Korea
e-mail: wjkim@glonik.com

R. Tutsch et al. (eds.), *Progress in Optomechatronic Technologies*,
Lecture Notes in Electrical Engineering 306, DOI: 10.1007/978-3-319-05711-8_9,

On the other hand, however, viewing from the structural aspect, regardless of either the early stage or modernized advanced microscopes, or of either low-priced or high-priced ones, they all consist of a series of objective lens and eyepiece, and further they are more or less similar in the respect that these two segments are fixed.

The observer sees the image of an object with 3-dimensional shape which is projected over the 2-dimensional plane, which signifies loss of 3-dimensional information of an object. In the past, either using special tilt stage or adaptor, or by moving the microscope itself (i.e., the user), the observation has been made for the object in a three-dimensional way although it is limited. But, whenever the observation angle changes, according to which the position of object also should change, so it is inconvenient. And, in such cases as when the observing object is heavy or substantially small and fragile, or when handling human or animals, or when observing flowing objects like liquid or gel etc., or when observing object or observation itself and operations are made simultaneously, it is virtually impossible to change the view of observation.

Recently, as the performance of camera improves with the development of digital technology, such products that enable to observe the object at oblique angles by moving camera itself are on the market [1–8]. However, they do not meet the satisfactions in terms of overall system size, pricing and level of image quality etc., further its usage is also limited in the field where simultaneous observation and operations are required, as it does not provide an environment of good hand-eye coordination.

In this paper, we propose the system that acquires the 3-dimensional information of an object such as height, step and angle etc., through use of the patent [9, 10] which is registered in Korea Institute of Patent and Information.

The methods of designing the optical mechanical parts and control system are separately explained in Chaps. 2, and 3. Chapter 4 will list the experimental results on the optics and controls, and drawing the conclusion is made in Chap. 5.

9.2 Optical and Mechanical Design

In this chapter, we show the design results of optical and mechanical parts of the system.

9.2.1 Designing Optics of the System

The optics of the system consists of a variety of optical elements such as lens, mirror and prism etc., in order to transfer images to the fixed position from moving objective lens. Figure 9.1 shows the optical ray paths of designed objective system. The optics is designed as much as symmetric to minimize certain types of aberrations with as fewer optical elements as possible. It also contains mirrors that transfers images of moving objective lens, and pechan prism that compensates the image rotation caused by rotation of mirrors. In addition, the section in which

Fig. 9.1 Image views of objective optic field

length changes over the optical path are designed to form infinity space to compensate the length change. In this example, the magnification of objective system is 1X, and the user magnifies the images formed by the objective lens using eyepiece. The magnification of eyepiece normally used ranges 10X–20X, and if users need various range of magnifications, the objective system should be redesigned or, more preferably, made zoom lens system.

9.2.2 Optical Performance

Various indices are involved in evaluating the performance of the optics system. Among them, the graph shown in Fig. 9.2 is representing the MTF (Modulation Transfer Function) that relates to resolution and visibility.

The Fig. 9.2 shows the MTF graph when observing perpendicular plane, and Fig. 9.3 indicates the graph when inclined 30°, i.e., when the polar angle equals to 30. There is no optical performance variation when moving in the direction of azimuth angle, since the whole objective system is rotating.

Without regard to the value of polar angle, modulation value at 30 cycles/mm of overall field area with 20 mm diameter shows good result with contrast ratio more than 0.5. With exception of peripheral field, it shows the performance approaching

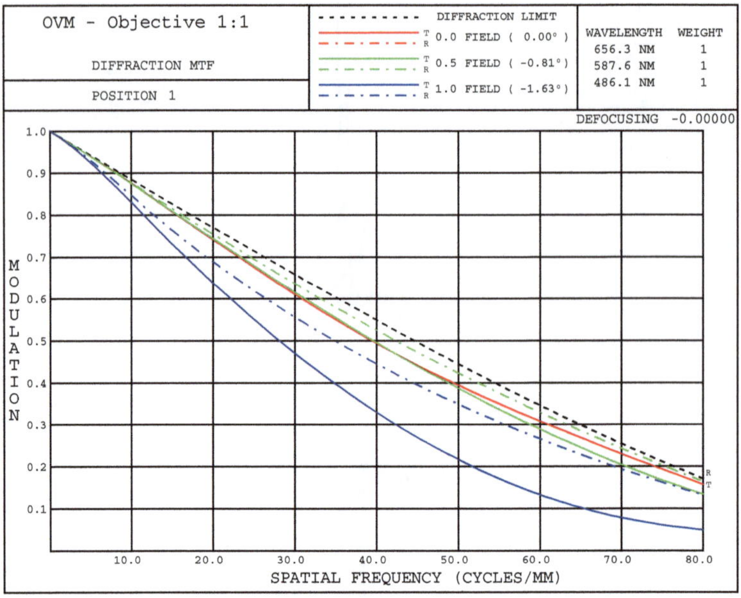

Fig. 9.2 Perpendicular plane observation MTF

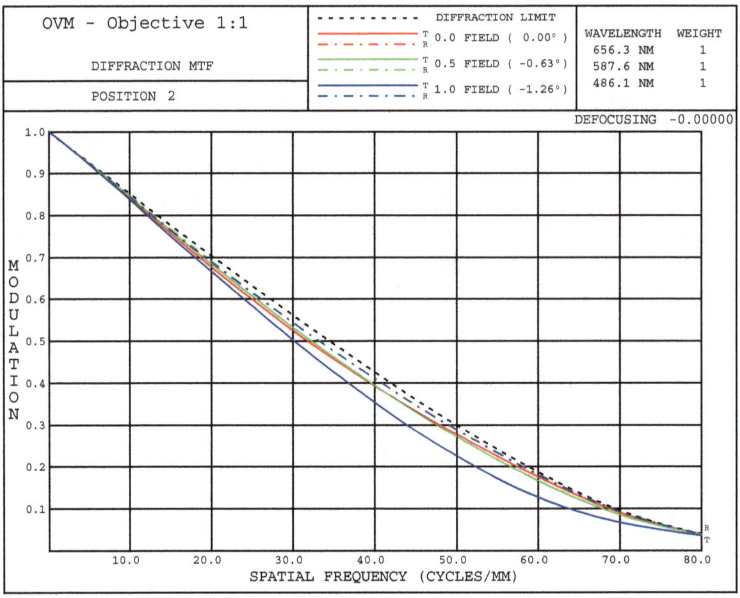

Fig. 9.3 Observation MTF inclined 30°

to the diffraction limit. Here we should also note that when inclined 30°, as the optical path length becomes lengthier, so the MTF value at diffraction limit becomes smaller.

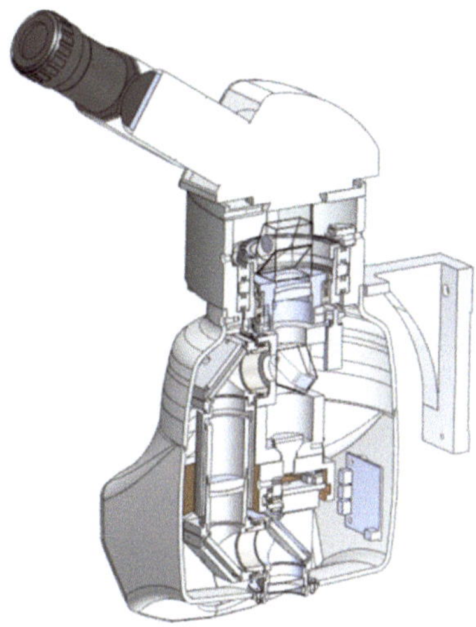

Fig. 9.4 Cross section of designed mechanical segment

9.2.3 *Designing Mechanical Segment of the System*

Figure 9.4 is the cross section of designed mechanical segment. Different from the conventional microscopes, the microscope used in the experiment for this paper has many moving parts. Thus more emphasis should be put on precision machining of each part so as to meet the tolerance range of the designed optical system. In particular, in order that the system reliability should be guaranteed, the optic axis should be kept aligned even after long term operations. To minimize torque when moving the polar angle and azimuth angle, bearings have been used where rotating or driving is concerned. In particular, the double row bearings have been all used except azimuth rotating area, thus minimizing angular deviation relative to the axial direction. By doing this, it ensures stable drive with even smaller than 1 W sized geared motor.

The Fig. 9.5 shows conceptual illustration made by 3D microscopic observation, which represents function when observing the object from arbitrary angle while objective lens is moving in the direction of polar angle and azimuth angle.

The Fig. 9.6 is the picture of its entire system.

9.2.4 *Principle of Measuring Steps*

This chapter introduces the method of measuring 3-dimensional information, i.e., measuring the depth information using 3D Microscope as described above. User properly establishes the polar angle and azimuth angle, and acquires angular value through sensor. Once image is acquired from the established position, it comes to enable calculating the 3-dimensional information through values of polar and azimuth angles as well as the values measured at image. Figure 9.7 shows principle of measuring steps of this system. This example illustrates that the step height can be calculated with the simple formula (9.1), once value of

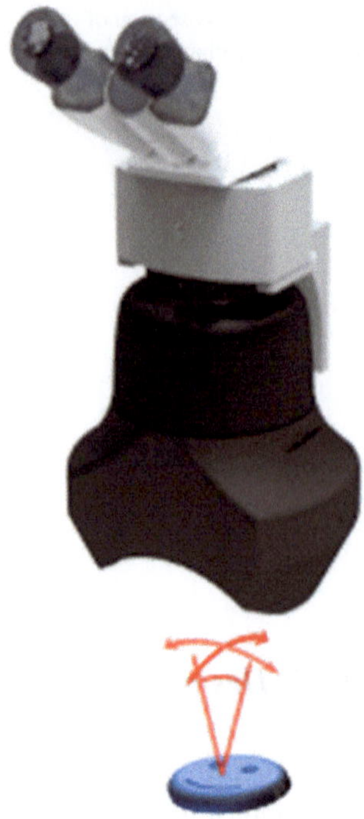

Fig. 9.5 Conceptual illustration by microscopic observation

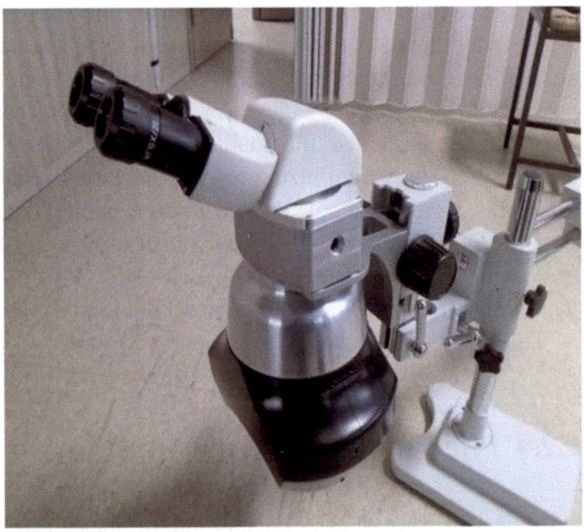

Fig. 9.6 Total developed system

polar angle and value of length at 2-dimensional image obtained from side plane angle are known.

$$h = \frac{p}{\sin \theta} \tag{9.1}$$

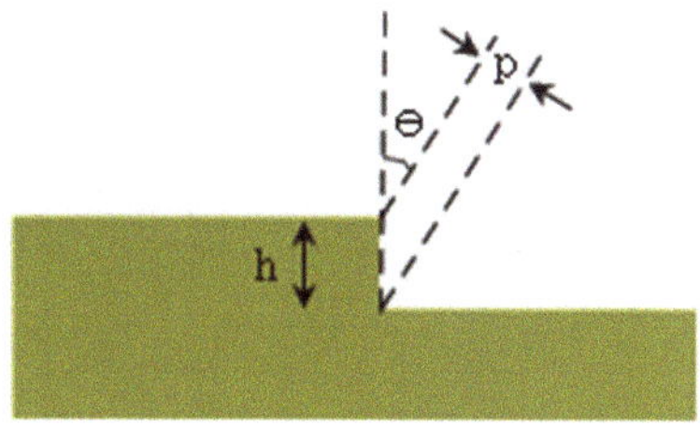

Fig. 9.7 Principle of measuring steps

This approach is not just much convenient and precise when compared with either the mechanical method which uses the existing various gages or the image processing method of z-axis scanning through acquiring multiple images in the direction of depth, but also can be applicable almost without limitation to the cases where measurements cannot be taken by the existing method.

Using formula (9.1), the step "h" can be simply obtained.

9.3 Designing Controller of the System

The Controller board has been designed using ATXmega16D4, and DC Motor is controlled using the driver chip BD6210F. Figure 9.8 is the block diagram of the Controller board.

As a way to independently move the fixed eyepiece and rotating objective lens at the mechanical segment, the Main Controller and Sub Controller are divided.

9.3.1 Main Controller

The Main Controller controls the azimuth motor of the eyepiece, and communicates with PC and Sub Controller. Its main function is receiving the Encoder data, and sending the Encoder data of the Main Controller and Sub Controller to PC. Also, making output of the received date onto the character LCD, from which enabling to identify the amplitude of the polar angle and azimuth angle, and adjusts the brightness of LED using the rotary switch.

9.3.2 Sub Controller

The Sub Controller controls both the polar angle motor at the objective lens segment and pechan prism motor that is used as image derotator. Using the data table already calculated for the rotated image which corresponds to the polar

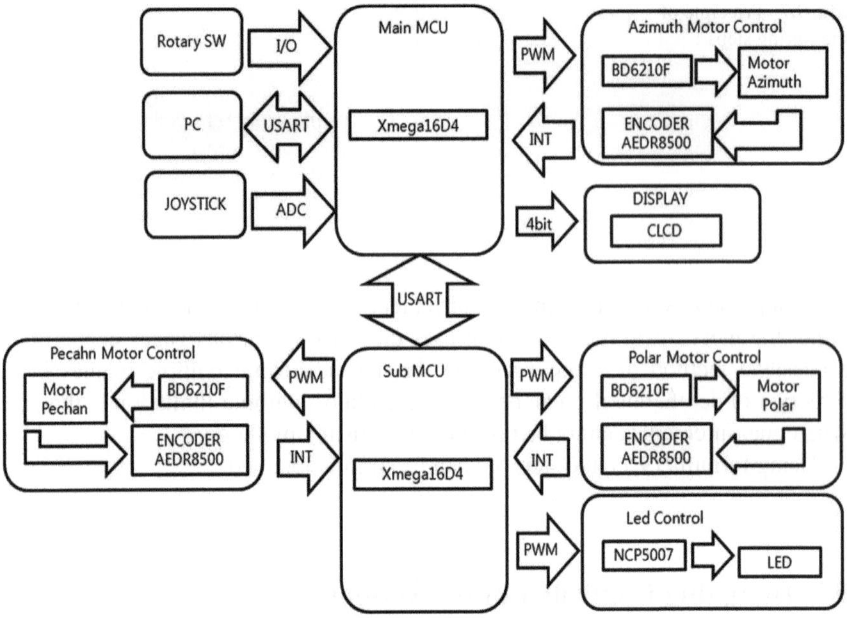

Fig. 9.8 Control system block diagram

angle, it synchronizes in real time the rotated image for its compensation. Using the LED driver chip NCP5007, it receives the value of rotary switch, thus adjusting LED brightness.

9.3.3 Optical Incremental Encoder

The value of polar angle is obtained using the Optical Incremental Encoder with 84 μm degree of precision. The mechanical design in this paper corresponds to approximately 0.03° if converted to the angular precision of the polar angle. For more precise control, signal can be multiplied 4 times.

The sensor adopted in this Controller is AEDR-8500, which is 3 channel reflective type optical encoder. Once set with absolute positioning value at power-on or reset, and it transfers the pulse signal to the Controller for measurements.

9.3.4 Operation Angle Motor

Polar and azimuth angle motors are controlled from the ADC values of joystick input. To get a good hand-eye coordination, main controller translates joystick movement into angle command of each motor. Figure 9.9 shows the block diagram of the drivers of polar and azimuth angle motors.

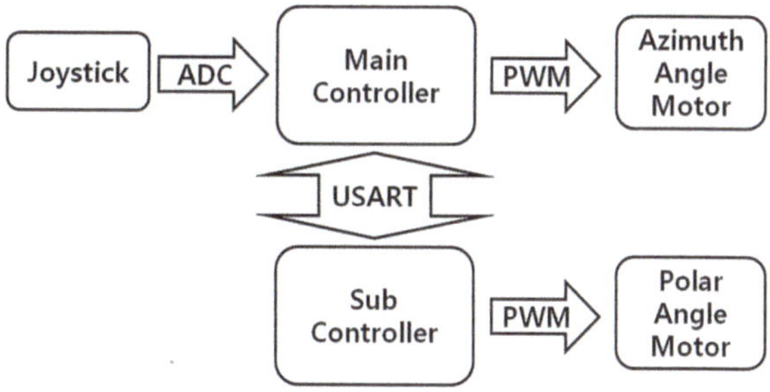

Fig. 9.9 Block diagram of motor drivers

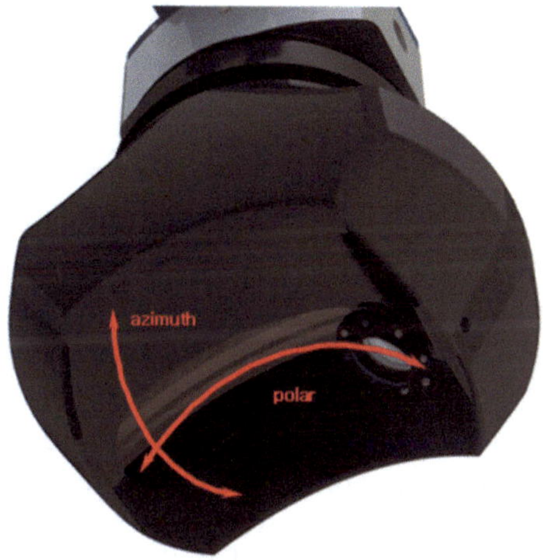

Fig. 9.10 Movement in azimuth and polar angle

Figure 9.10 shows the movement in azimuth angle and polar angle. The azimuth angle motor rotates 360° without limit while the polar angle motor moves within a pre-defined angle around the observation object.

9.4 Measurement Result

This chapter notes the actual measurements that have been implemented using the manufactured system with the elements of the optics and mechanical segment, drive and controller, and software etc.

9.4.1 Drive and Control

In order to compensate the image rotation, the pechan prism should be rotated relative to the polar angle movement of the objective lens. This function of image derotator is controlled by the Servo System using small sized DC Motor and Optical Encoder.

Fig. 9.11 Extracted capacitor image

9.4.2 Integrated Test Results of Optics and Control

The test results of the Controller have been made identifying the value with image extracted by Software. Using the camera of Moticam3000 of Motic Corp., the image has been transferred to PC, and the software for PC has been developed with MFC using SDK of Motic Corp.

For convenience purpose of measurements, it has been included with settings of rgb gain value, offset, resolution and exposure time etc. In addition, included is the joystick function, by which to enable not only to drive the Controller but also to drive polar angle and azimuth angle with software.

The Fig. 9.11 is image actually acquired using Moticam3000. With the scene observed from the perpendicular plane, image of SMD type capacitor was captured.

With the above image, only about the upper plane distance of the soldered capacitor can be obtained. The Fig. 9.12 is the image of inclined 27.5° of the polar angle.

As in above image, the side plane of SMD type capacitor can be observed. Using formula (9.1), "h" can be calculated. The following Fig. 9.13 is the image that captured the scene actually measured

Figure 9.14 is the measured image of SMD type capacitor using software.

The software has been calibrated over the program. Using Encoder value, the polar angle value is known, and as the side length has been measured 0.58 mm,

Fig. 9.12 Capacitor image inclined 27.5°

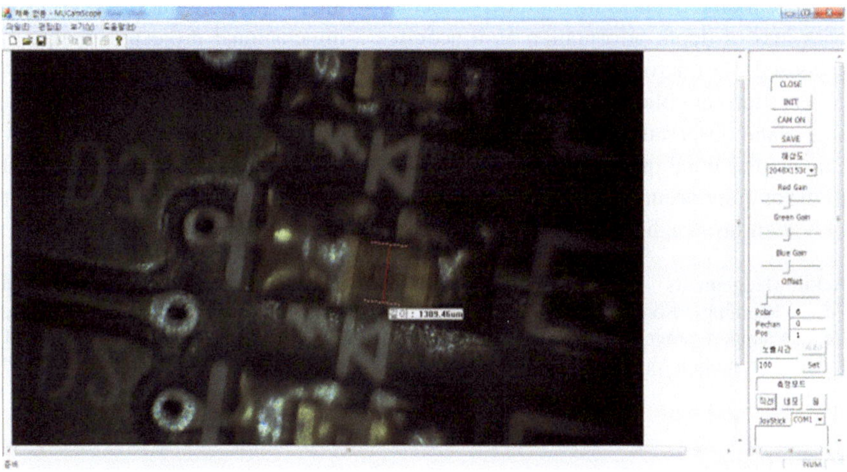

Fig. 9.13 Measured scene using software

$1.26 = \frac{0.58}{\sin 27.5}$ is obtained using the formula (9.1). The real height of SMD type capacitor measures 1.25 mm, showing approximately 1 % error rate, resulting in much precise measurement result.

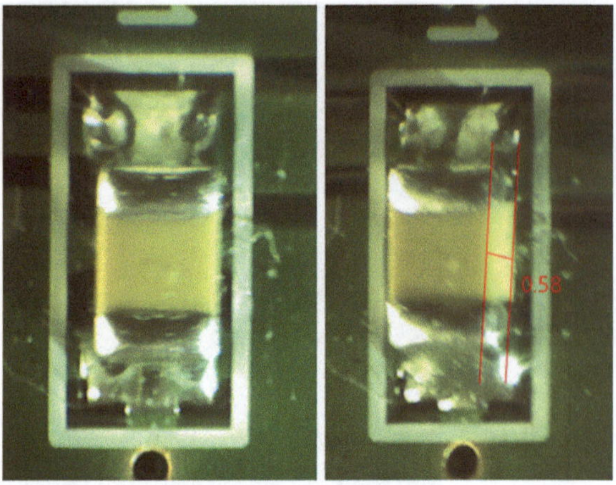

Fig. 9.14 Step measurement

9.5 Conclusion

This paper proposes the system that acquires and measures the 3-dimensional information of the object such as height, depth, step, angle etc., and has represented its system effectiveness through experiments. While the existing microscope ensues with loss of depth information as it acquires 2-dimenstional projected image, the proposed 3D microscope enables to move the objective lens, thus facilitating the observations at oblique angles, meaning not only observing perpendicular plane but side plane observation as well. Therefore, when using 3D microscope, once if we can acquire good quality image through optimal illumination, it becomes possible to make quite accurate measurement of 3-dimensional features such as step, depth of hole or groove, height of protruded part etc. with simple operations.

Acknowledgements This research was supported by the MSIP (Ministry of Science, ICT and Future Planning), Korea, under the C-ITRC (Convergence Information Technology Research Center) support program (NIPA-2013-H0401-13-1006) supervised by the NIPA (National IT Industry Promotion Agency)

References

1. Tao X, Cho H, Janabi-Sharifi F (2008) Active optical system for variable view imaging of micro objects with emphasis on kinematic analysis. Appl Opt 47(22):4121–4132
2. Potsaid B, Bellouard Y, Wen J (2005) Adaptive scanning optical microscope (ASOM): a multidisciplinary optical microscope design for large field of view and high resolution imaging. Opt Express 13(17):6504–6518
3. Friedlander T, Kajiro Y (2007) Apparatus for observing protein crystals. The United States Patent and Trademark Office, US20070217007 A1
4. The Secretary of State for Defence (2008) Viewing apparatus. Korea Institute of Patent Information, 10-2008-7031764
5. DongWon EnC (2008) Microscope for photographing object in many direction. Korea Institute of Patent Information, 20-2008-0021782

6. Amarel Precision Instruments, Inc (1993) Oblique viewing system for microscopes. The United States Patent and Trademark Office, US5253106 A
7. Lei Li, Allen Y (2010) Design and fabrication of a freeform prism array for 3D microscopy. JOSA A 27(12):2613–2620
8. Olympus. Opto-digital Microscope, http://www.olympus-ims.com/ko/microscope/dsx100
9. Kim WJ, Jung DH (2013) Optical system and method for extracting three dimension information using the same. Application for a Korea Institute of Patent Information, 10-2013-0009727
10. Kim WJ (2011) Optical system for forming optical path of oblique angle and method thereof. Korea Institute of Patent Information, 10-2011-0006799

Part III
Optical-Based Mechanical Systems Including Sensors and Actuators

Chapter 10
Design and Fabrication of a Prototype Actuator for Fourier Transform Interferometry

G. G. Melpignano, A. Schaap and Y. Bellouard

Abstract The design, fabrication and characterisation of an uniaxial scanning-mirror component for interferometry applications is presented. The device is fabricated in fused silica by means of femtosecond laser irradiation and a selective etching step. The fabricated actuator produces bidirectional, repeatable sinusoidal motion with a displacement (200 μm) larger than the wavelengths under investigation (300–700 nm), rendering it suitable for application within an interferometry setup. This indicates that future fabrication of fully integrated interferometer measurement devices in fused silica are realisable.

Keywords Fourier transform spectrometry • Fused silica • Femtosecond laser machining • Halbach magnet array • Actuator • Birefringence • Microsystems

10.1 Introduction

Fourier Transform Spectrometry (FTS) is a well-known spectrometry method which uses the principle of interferometry to obtain absorption and/or emission spectra. However, due to the requirement of a highly accurate mirror-scanning component, the current FTS devices are large in size at high cost.

The main goal of this investigation is to design, fabricate and characterise a proof-of-concept for a miniaturised mirror-scanning component which could function within an interferometry setup, fabricated by means of femtosecond laser irradiation in fused silica glass.

Over the recent years, the miniaturisation of FTS devices in silicon has been the focus of multiple research groups [1–4]. The integration of the created devices in combination with other functional components, e.g. micro-fluidic channels, has yet to be shown. This results from the steep increase in the manufacturing process

G. G. Melpignano (✉) · A. Schaap · Y. Bellouard
Eindhoven University of Technology, Den Dolech 2, 5612 AZ Eindhoven, The Netherlands
e-mail: g.melpignano@student.tue.nl

R. Tutsch et al. (eds.), *Progress in Optomechatronic Technologies*,
Lecture Notes in Electrical Engineering 306, DOI: 10.1007/978-3-319-05711-8_10,

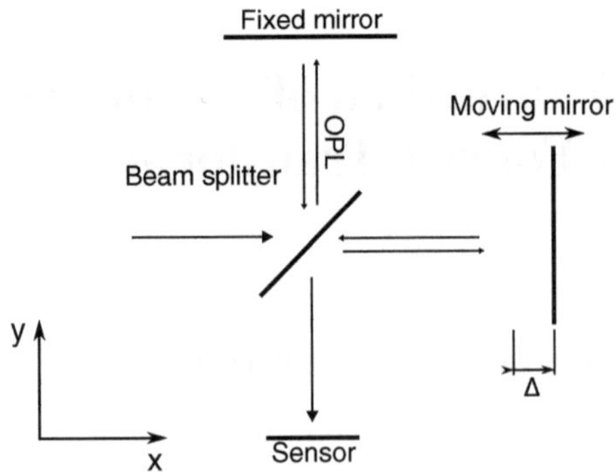

Fig. 10.1 Schematic representation of a Fourier Transform Spectrometer (FTS) setup which can be used for the measurement of an absorption spectrum. Incident light is redirected by a beam-splitter to reflect upon two mirrors. Due to the displacement of one of the mirrors, a phase shift is introduced. Recombination at the beamsplitter of the reflected light causes constructive or deconstructive interference. The intensity of the light is measured with a sensor

due to the anisotropy of the crystalline material, constraining designs to 2, 5D. Additionally, silicon is non-transparent to visible wavelengths which impedes the optical properties of the device.

Femtosecond laser machining has been shown to be a versatile technique capable of producing structures with different functionalities in fused silica glass. Fused silica glass is an amorphous material with an isotropic material behaviour. By irradiating the fused silica with ultrashort, high power pulses, the material at the focal spot of the laser is locally modified. This causes preferential etching behavior of the modified glass [5]. The process enables the manufacturing of complex 3D structures. The high flexibility of manufacturing different classes of systems thus enables the production of multi-functional lab-on-a-chip devices [6].

A schematic representation of an FTS measurement setup is shown in Fig. 10.1. The FTS is built up out of four components; (1) Beamsplitter; (2) Fixed mirror; (3) Scanning mirror; (4) Sensor. The setup relies upon creating differences in optical path length (OPL) between light going toward and coming from the stationary and moving scanning mirror. The light from the two mirrors is recombined at the beamsplitter and directed towards the sensor. Due to the displacement Δ, these beams will interfere with one another. Interference can range from constructive to destructive, with all the possibilities in between.

The constructive and deconstructive interference for an equal displacement, Δ, varies for each wavelength, as the OPL difference is coupled to the wavelength [7]. Therefore, within the same displacement, wavelengths of shorter length will construct or deconstruct more often than wavelengths of larger lengths.

The sensor will measure the total intensity contribution of all electromagnetic waves over time. As the mirror displaces, the intensity will fluctuate. A Fast Fourier Transformation (FFT) of the intensity over time will result in its equivalent frequency spectrum of the electromagnetic waves under investigation.

The requirements of the device to be fabricated are; (1) one-directional side ways motion in the XY-plane, (2) a surface perpendicular to the direction of motion which could act as a mirroring surface, (3) reproducible motion over time, (4) large displacements (>100 μm) with respect to the electromagnetic spectrum under investigation (300–700 nm).

10.2 Design and Fabrication

The mirror scanning component will consist of a moving mechanism in combination with an actuator. The mechanism will limit the motion of the mirror-scanning component in order to meet the defined requirements, whilst the actuation principle produces forces to displace the mechanism.

The mechanism is a monolithic double compound linear guidance mechanism (LGM) which uses flexure leaf springs in order to produce bi-lateral reproducible motion in the $\vec{x}$-direction over time, see Fig. 10.2a. The mechanism relies upon the elastic deformation of the flexure leaf springs in order to assure reproducibility. The effective stiffness of the mechanism, for small deformations which assume linear elasticity, is given as;

$$K = \frac{2Ebh^3}{L^3} \tag{10.1}$$

with E the Young's modulus of the fused silica and b, h and L the height, width and length of the flexure leaf springs which interconnect the bodies [14].

The design is completely symmetric resulting in better balancing, a symmetric load distribution, expansion properties under thermal loading and a diagonal stiffness matrix. Additionally, each of the flexure leaf springs displaces half of the total displacement required from the main moving body [8].

The dimensioning of the LGM is limited by the dimensions of the fused silica slides (23 × 23 × 0.5 mm), fabrication method, stiffness requirements, dimensioning of the available parts and its handling capabilities for manual assembly. The flexure leaf springs determine the stiffness of the mechanism. Each spring has been designed with a width, height and length of 25 μm × 500 μm × 8 mm.

Actuation is applied in the form of electromagnetic actuation. In the case where there is no electric field within the conductor, the forces are given as:

$$\boldsymbol{F} = \boldsymbol{B} \times \boldsymbol{J} \cdot V \tag{10.2}$$

where $\boldsymbol{B}$ the magnetic flux density produced by a magnetic setup, $\boldsymbol{J}$ is the current density through and V is the volume of the current carrier respectively.

The magnetic field is produced by an Halbach array shown in Fig. 10.2b [9, 10]. By placing the magnets in the configuration shown in Fig. 10.2b, the magnetic field is strengthened on one side of the magnet array and decreased on the other, resulting in a more efficient use of the magnetic field. A single turn current carrier is placed within the magnetic field, which results in the production of

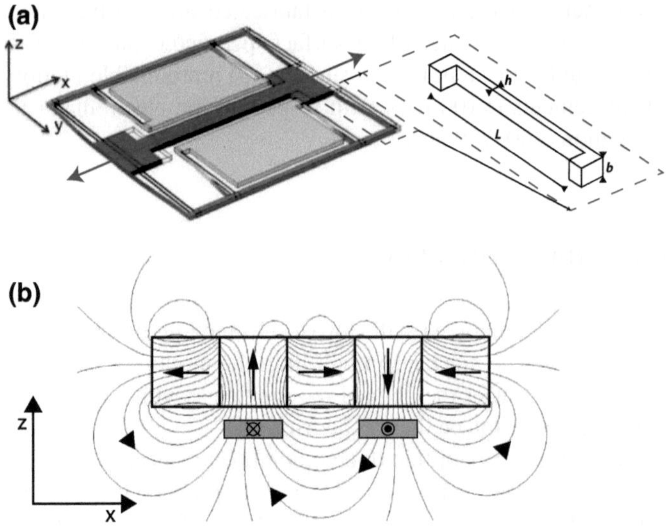

Fig. 10.2 **a** Design of the double compound linear guidance mechanism. In *green*, the static bodies of the mechanism are shown. In *blue*, the moving components of the device are shown. **b** Halbach magnet array which due to its particular magnet configuration produces a strong one-sided magnetic field. By placing current carrying components within the magnetic fields, a electromagnetic force is produced in the $\vec{x}$-direction. The magnet array is placed within the largest moving component of the linear guidance mechanism

a force in the $\vec{x}$-direction. Commercially available 1 mm cubic magnets are used for this setup. The width, height and lengths of the single turn coil are 1, 0.1 and 40 mm respectively.

The proof-of-concept device will consist of monolithic double compound linear guidance mechanism containing the magnet array, stacked on a second fused silica slide with a copper carrying component embedded into the slide. The gap distance, 0.2 mm, between the magnet array and current carrier is ensured by the application of spacers. In order to ensure proper alignment of the layers with respect to each other, the individual components are slid over allignment pins (Fig. 10.3).

The bottom slide with the copper single turn coil, and the LGM mechanism with the magnet array, are fabricated by means of the femto-second machining process in fused silica. The exposure step is done with 380 fs-long pulses emitted at 1030 nm from a diode pumped Ytterbium-KGW crystal based femtosecond oscillator (t-Pulse 500 from Amplitude Systemes). The oscillator emits pulses at a frequency of 9.4 MHz. An acousto-optic modulator (AOM) is used to reduce the repetition rate by chopping out pulses from the main pulse train.

For fabricating this device, a repetition rate of 400 kHz was used. The laser beam is focused using a 20× objective with a numerical aperture (NA) of 0.40. Using three linear positioning stages, the specimen is moved with an accuracy of

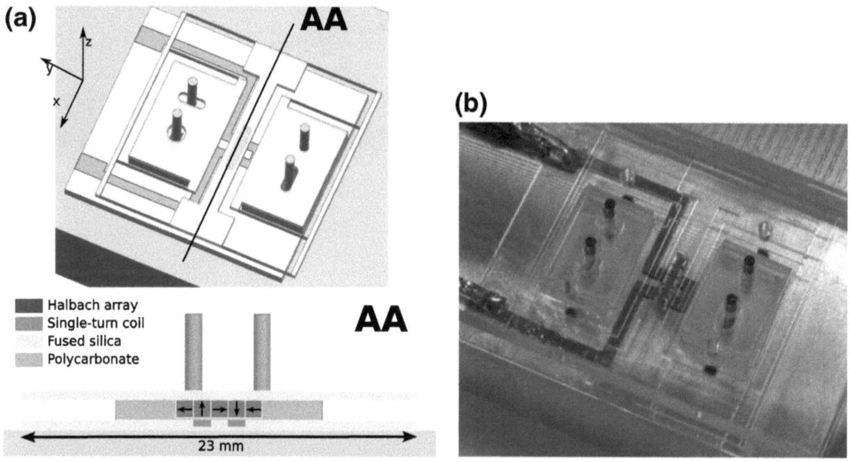

Fig. 10.3 **a** Designed double compound linear guidance mechanism with the implemented magnetic array which is stacked on a second slide containing a single turn coil. By application of a current, electromagnetic forces are created, which cause motion of the guidance mechanism. **b** Completed fabricated device

1 μm under the objective. The writing speed is 2 mm/s and the pulse energy is 225 nJ. The second process step is a chemical etching step. The laser modified-substrate is immersed into a low concentration aqueous HF bath (2.5 %) for 40 h. For the fabrication of the LGM it was chosen to machine wider leaf springs and leave them in the etchant for a larger period of time. This decreases the possibility of imperfections on the surface area, resulting in higher loading capabilities of the flexures. The initial flexure leaf spring width was set to 100 μm [11].

10.3 Experiments

The device should be capable of producing reproducible motion over time in order to assure proper functioning within a complete measurement setup. The response of the device to sinusoidal input currents is investigated by measuring the displacement of the main moving body of the mechanism.

The experimental setup consists of a E/box 1.05 signal generator connected to a LDA PA25E shaker amplifier. The output displacement of the device is measured by a laser triangulation sensor, Keyence LK-G10, and read-out using the data acquisition ports of the E/box 1.05. An input chirp signal has been applied with a Hamming window, ranging from 10–200 Hz in 10 s.

In order to verify the experiments, a multiple degree-of-freedom model has been created for which each of the bodies only exert unidirectional motion in $\vec{x}$. Assuming mass m_i of the intermediate bodies and main body, stiffness k_i of

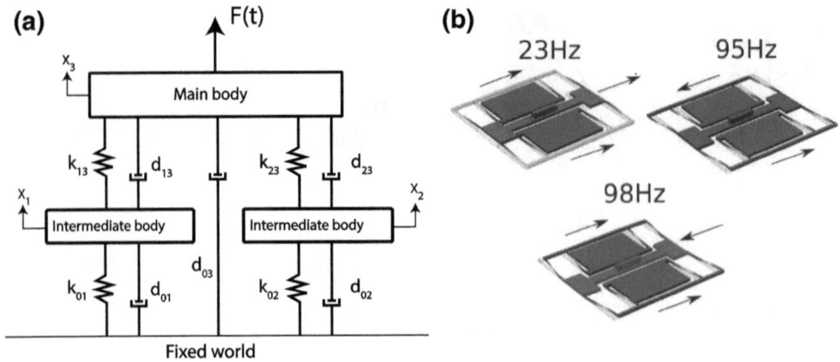

Fig. 10.4 **a** Simplified model of the guidance mechanism used for the simulation of the devices behaviour. **b** First three natural frequencies of the device in its vibrational modes

two parallel leaf springs, damping k_i due to friction, air damping and magnetic damping; the equations of motion for the three masses shown in Fig. 10.4a:

$$\underline{M}\ddot{\underline{q}}(t) + \underline{D}\dot{\underline{q}}(t) + \underline{K}\underline{q}(t) = \underline{Q}(t) \tag{10.3}$$

$$\underline{q}(t) = \begin{pmatrix} x_1 \\ x_2 \\ x_3 \end{pmatrix} \quad \underline{Q}(t) = \begin{pmatrix} 0 \\ 0 \\ F(t) \end{pmatrix} \quad \underline{M} = \begin{pmatrix} m_1 & 0 & 0 \\ 0 & m_2 & 0 \\ 0 & 0 & m_3 \end{pmatrix}$$

$$\underline{K} = \begin{pmatrix} k_{01} + k_{13} & 0 & -k_{13} \\ 0 & k_{02} + k_{23} & -k_{23} \\ -k_{13} & -k_{23} & k_{13} + k_{23} \end{pmatrix}$$

$$\underline{D} = \begin{pmatrix} d_{01} + d_{13} & 0 & -d_{13} \\ 0 & d_{02} + d_{23} & -d_{23} \\ -d_{13} & -d_{23} & d_{13} + d_{23} + d_{03} \end{pmatrix}$$

Masses have been calculated from the dimensions and properties of materials used. In the current model, the mass of the individual flexure leaf springs is neglected. The stiffness of the device has been extracted from Eq. 10.1. Minimal amounts of air damping, 0.1 %, have been assumed in order to smoothen the response of the simulations.The first three eigenfrequencies are shown in Fig. 10.4b.

As the guidance mechanism is made in fused silica, it is possible to conduct birefringence measurements to investigate the stresses within the material. Bellouard et al. [11] have found stresses as high as 1.3 GPa within a single bending fused silica flexure. Investigating the stresses gives an indication of the present load within the flexure leaf springs.

By means of a dark-field LC Polscope polariscope, the relative angular phase shift in the material is measured. A dark-field polariscope consists of light passing through a polariser, onto a quarter-wave plate, through the sample under investigation, onto a second quarter-wave plate and finally through a polariser placed at 90°

with respect to the first polariser. The intensity of the light for a constant angle between the polariser and birefringence axis can be expressed as [11]:

$$I(x, y) = \sin^2 \left(\frac{\theta(x, y)}{2} \right) \tag{10.4}$$

with the angular phase shift, θ, resulting from the main principle stresses, σ_i, as:

$$\theta = 2\pi C(\sigma_y - \sigma_x)\frac{t}{\lambda_0} \tag{10.5}$$

where $C = 3.55 \cdot 10^{-12}\,\mathrm{Pa}^{-1}$ is the Brewster constant of fused silica, $t = 460\,\mu\mathrm{m}$ the thickness of the specimen and $\lambda_0 = 546$ nm the wavelength of the light source used for the investigation. Interference fringes will occur for integer values of m when $\theta = m2\pi$, see Eq. 10.4. The stress can be expressed as:

$$(\sigma_y - \sigma_x) = m\frac{\lambda_0}{Ct} \tag{10.6}$$

10.4 Discussion

The sinusoidal response of the device is shown in Fig. 10.5a. The displacement of the main body reproduces the sinusoidal input signal as a displacement, indicating reproducibility of motion over time. The force response of the device for a sweep of frequencies is shown in Fig. 10.5b. Differences between measurements and simulations results from inequalities in the width of the flexure leaf springs. As the stiffness of the device depends cubically on the width of each of the flexure leaf springs, small irregularities can have large influences on the functioning of the device. Irregularities between flexure leaf springs cause a loss of symmetry, resulting in possible additional rotations and translations occurring in the motion of the device.

In this case the imperfections cause an increase in the stiffness of the completed device, resulting in an increased first natural frequency. The second and third natural frequency could not be verified due to the decreasing coherence. The irregular phase jumps of 360° between 40–80 Hz result from the software which has been used to determine the transfer function of the device from the measurement data.

Additionally, Joule heating has been evaluated by means of measuring resistance fluctuations of the copper coil. These were too small to be measured with the equipment used. Therefore, the device showed the capability of conducting current densities of $10^7\,\mathrm{Am}^{-2}$.

The birefringence measurements, see Fig. 10.6, show that the deformation in the flexure leaf springs is in agreement with the linear elastic theory. A neutral axis lies in the middle, with an increasing symmetric stress distribution over the width of the leaf spring. Upon bending, the material shows elongation and compression on opposite sides of the neutral axis. This results in equal amounts of birefringence, resulting in equal intensities. It should be noticed in Fig.10.6 that the intensity distribution represents the absolute stress distribution within the material. On average, for 16 measurement points on 8 flexure leaf springs, the average stress

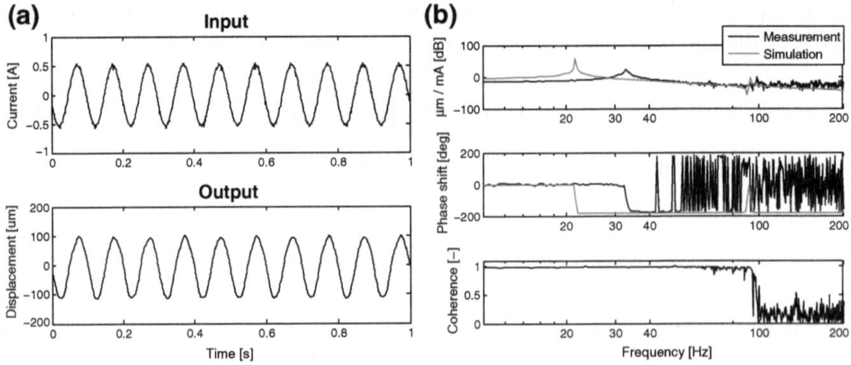

Fig. 10.5 **a** Sinusoidal displacement response of the device due to the application of a sinusoidal input current. **b** Amplitude, phase and coherence of the forced response measurements of the device for an input current and output displacement

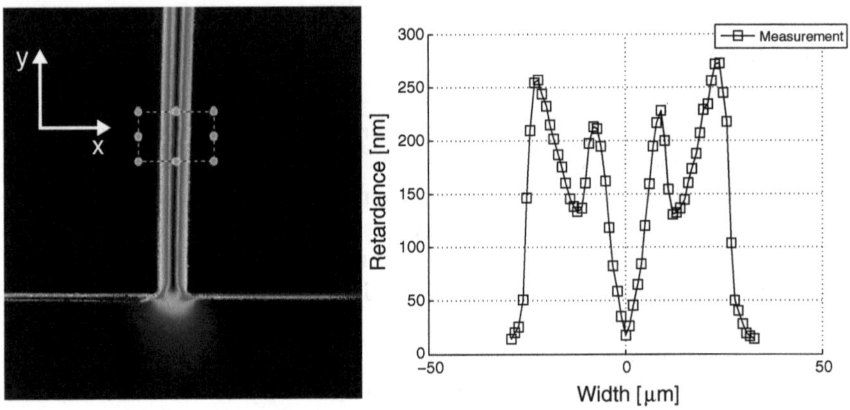

Fig. 10.6 *Left* Grayscale picture obtained from the LC Polscope measurement setup of one of the flexure leaf springs connected to one of the main bodies. The presence of the neutral axis line can be seen, as expected from linear elastic theory. For an increased deformation, lighter regions appear. The *box* indicates the section over which the cross-sectional intensity distribution has been averaged in the $\vec{y}$-direction. *Right* Cross-sectional average of the intensity distribution over the width of the flexure leaf spring. The results shown 1.5 distinct peaks. Therefore $m = 1.5$ in Eq. 10.4, resulting in an average stress of about ~550 MPa

was about 500–700 MPa within the flexure leaf springs, depending on the width of the leaf spring. The displacement of the device could therefore be increased, as the stress remains well below the previously reported 1.3 GPa [11].

The partially transparent actuator can be used in various optical applications such as adaptive optics and microscopy. Thanks to the broad transparent spectral range of fused silica, transparent actuators can be used for creating tuneable resonant cavities in the visible spectrum [12]. The transparency of the current actuator could be increased by reducing the device to a single layer and using transparent magnets and coils. A transparent coil of indium-tin-oxide could be created in a fused silica cavity. Transparent magnets are currently being researched in the work of Gich et al. [13], where magnetic silica-aerogel composites have been synthesised using a fast sol-gel process and subsequently super-critically drying the resulting gels.

10.5 Conclusion

A successful proof-of-concept partially-transparent actuator has been fabricated by means of femtosecond laser machining in fused silica. This demonstrates the possibility of integrating a scanning mirror component for an interferometry setup. Thus, fully integrated interferometer measurement devices in fused silica in combination with other functionalities (like waveguides, polarisation converter devices, bulk-laser fabricated micro-optical components, direct-write Fresnel lenses and bulk Bragg-gratings) can be realised in fused silica using femtosecond laser machining [6, 12].

References

1. Manzardo O, Herzig HP, Marxer CR, de Rooij NF (1999) Miniaturized time-scanning Fourier transform spectrometer based on silicon technology. Opt Lett 24:1705–1707
2. Montgomery PC, Montaner D, Manzardo O, Flury M, Herzig HP (2004) The metrology of a miniature FT spectrometer MOEMS device using white light scanning interference microscopy. Thin Solid Films 450:79–83
3. Knipp D, Stiebig H, Bhalotra SR, Bunte E, Kung HL, Miller DAB (2005) Silicon-based micro-Fourier spectrometer. IEEE Trans Electron Devices 52:419–426
4. Wallrabe U, Solf C, Mohr J, Korvink JG (2005) Miniaturised Fourier transform spectrometer for the near infrared wavelength regime incorporating an electromagnetic linear actuator. Sens. Actuators, A-Phys 123:459–467
5. Marcinkevicius A, Juodkazis S, Watanabe M, Miwa M, Matsuo S, Misawa H, Nishii J (2001) Femtosecond laser-assisted three-dimensional microfabrication in silica. Opt Lett 26:277–279
6. Bellouard Y, Champion A, Lenssen B, Matteucci M, Schaap A et al (2012) The Femtoprint project (Invited). J Laser Micro Nanomachining 7:1–10
7. Smith BC (1995) Fundamentals of Fourier transform infrared spectroscopy. Taylor & Francis, London
8. Jones RV (1951) Parallel and rectilinear spring movements. J Sci Instrum 28:38–42
9. Lee MG, Lee SQ, Gweon SD (2004) Analysis of Halbach magnet array and its application to linear motor. Mechatronics 28:115–128
10. Hao J, Xing P, Bai L (2004) Design and analysis of moving magnet synchronous surface motor with linear halbach array. Procedia Eng 28:115–128
11. Bellouard Y (2011) On the bending strength of fused silica flexures fabricated by ultrafast lasers. Opt Mater Express 1:816–831
12. Lenssen B, Bellouard Y (2012) Optically transparent glass micro-actuator fabricated by femtosecond laser exposure and chemical etching. Appl Phys Lett 101:103503
13. Gich M, Casas LI, Roig A, Molins E, Sort J et al (2003) High-coercivity ultralight transparent magnets. Appl Phys Lett 82:4807–4809
14. Bellouard Y (2010) Microrobotics: methods and application. CRC Press, Boca Raton

Chapter 11
Peak Search Method for the Optical Spectrum Signals

Riqing Lv and Yong Zhao

Abstract Nowadays, optical fiber sensors were widely used in physics and chemistry. And the processing of the optical spectrum signals was important in the sensing area. In the paper, a simple peak and notch search method was proposed to obtain the wavelengths of peaks and notches in the fiber Fabry-Perot interferometer sensor. The method efficiently reduced the noise of the optical spectrum and detected the wavelengths of peaks and notches, and it could be also extended for other optical spectrum signals.

Keywords Optical fiber sensor • Peak search method • Peak wavelength • Optical spectrum signals • Fiber Fabry-Perot interferometer sensor

11.1 Introduction

Now more and more fiber optical sensors were widely used in the sensing area for many physical and chemical parameters. The processing of the optical spectrum signals was one of the important procedures. A suitable peak-search routine was quite important to archive accuracy results. And there are many such methods: the peak was a sum of two features, the 'background' and the 'peak' [1, 2]. Using a nonlinear curving fitting function of several parameters to obtain an appropriate curving compared with original optical spectrum [3, 4], such as Gaussian fit, polynomial fit, cubic spline interpolant. The detection of the peak was based on the

R. Lv · Y. Zhao
College of Information Science and Engineering, Northeastern University, Shenyang, China

Y. Zhao (✉)
State Key Laboratory of Synthetical Automation for Process Industries, Northeastern University, Shenyang, China
e-mail: zhaoyong@ise.neu.edu.cn

R. Tutsch et al. (eds.), *Progress in Optomechatronic Technologies*,
Lecture Notes in Electrical Engineering 306, DOI: 10.1007/978-3-319-05711-8_11,

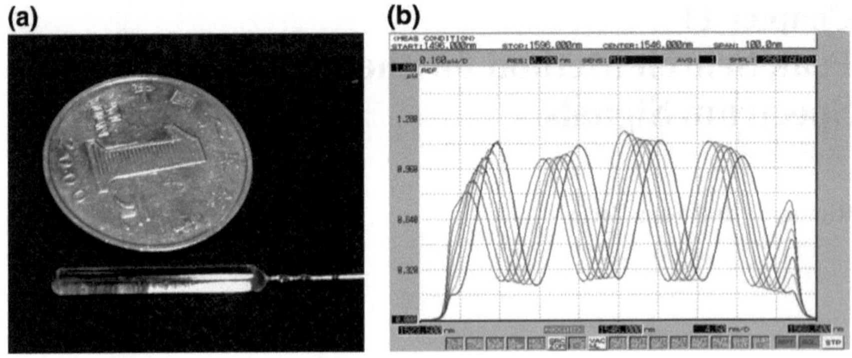

Fig. 11.1 **a** A Fabry-Perot sensor. **b** The spectrum of the sensor versus temperature

spectrum convolution [5, 6] and all stages of the peak-search processing used the convoluted spectrum instead of the original one, which was usually only used for peak location. However, it was difficult to find a proper function to curve fitting or convolution and it also wasted much time. And the detection of doublets and multi peaks were difficult to realize. Then these problems made that it is not easy to archive for the peak search with the methods. And there was another point that all these procedures were only to aim at peak searching but not fit to search the notch.

In the paper, a simple peak search method was proposed to mainly use in the fiber Fabry-Perot interferometer sensors here, it would not need much time for processing and it could search multi peaks and notches. It was mainly used in the fiber Fabry-Perot interferometer sensors here. And compared the spectrum of Fabry-Perot interferometer sensors and that of other fiber sensors, the simple peak search method was also quite fit to the peak search of the optical spectrum signals of most fiber optic sensors. Firstly, since the optical spectrum signals contains many noise, such as the electronic noise from the photodetector and relative circuits and the optical noise, so the noise reduction of the optical spectrum signals was an important part. Secondly, the most important procedure was the determination of the peaks and notches of the optical spectrum signals. And in the paper, the method was mainly used to search the peaks and notches of the optical spectrum signals.

11.2 Noise Reduction of Optical Spectrum Signals

In the optical spectrum signals, it contains many noises, such as electronic noise, optical noise and fabrication error. Before the peak search, the procedure of noise reduction should be done.

The optical spectrum of Fabry-Perot interferometer sensor with the 94 μm cavity length made by our own laboratory was shown as Fig. 11.1a. Its size was small compared with a coin of one Yuan. And the optical spectrum signals of the sensor

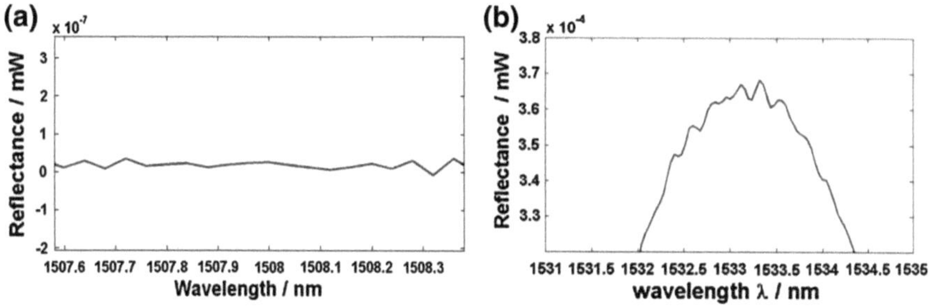

Fig. 11.2 **a** The baseline noise. **b** The optical noise

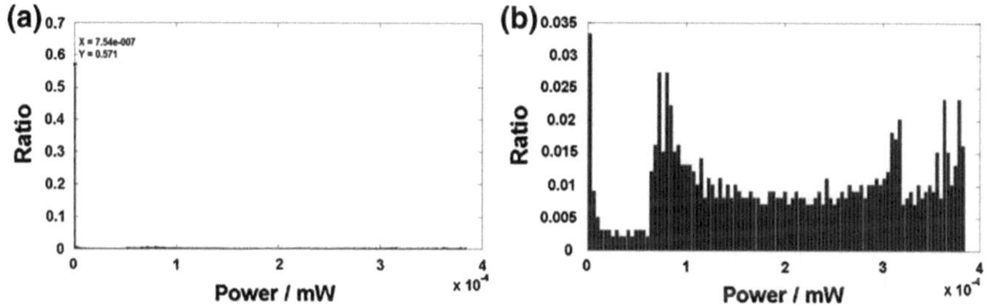

Fig. 11.3 The histograms of optical spectrum signals

measured with the increasing of temperature were also shown as Fig. 11.1b from optical spectrum analyzer (OSA).

The signals seemed to be good enough to get the right positions of the peaks and notches. But when zoom the spectrum signals in, there would be many details about the peaks as shown in Fig. 11.2. Figure 11.2a showed the baseline of the optical spectrum signals and it could obtain that the electronic noise from the photodetector and relative circuits. It maybe was not an important problem to detect the peaks and notches for other methods. In the paper, this noise was deal with by using histogram and a threshold because it hold a large part of the signals. The method of histogram was used to statistic the numbers of each power of the signals with the wavelength range from 1,520 to 1,570 nm as shown in Fig. 11.3a. It was found that the baseline of the optical spectrum signal was about 7.54×10^{-7} mW significantly.

Then the baseline data could be used to calculate the noise of dark current of the photodetector and the noise in Fig. 11.2a could be reduced by making the values of data beneath the baseline equal to zero. But such as the Fig. 11.3b, it would not influence the vital signals, so the baseline would not need any process.

And in the express of the procedure was as follow

$$\begin{cases} r_i = \frac{n_i}{N} \quad i = 1, 2, 3 \ldots m; \\ N = \sum_{1}^{m} n_i \end{cases} \tag{11.1}$$

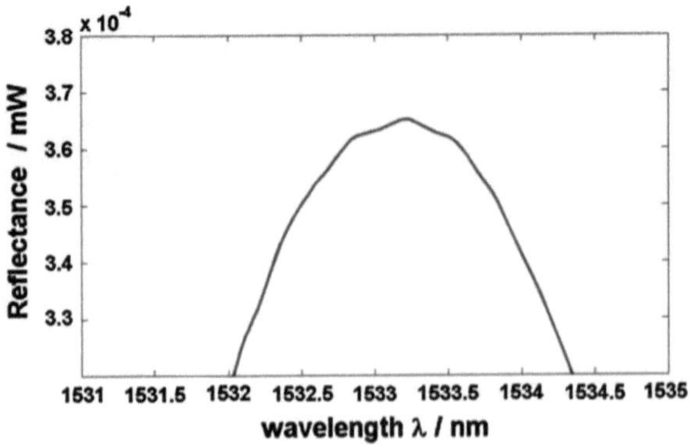

Fig. 11.4 The spectrum signals of Fig. 11.2b after noise removal

where, N is the total numbers of the signals, m is total groups of the signal equal to $N/10$, n_i is the numbers of the each group, r_i is the ratio of the each group. Then the histogram could be archived. And Eq. (11.2) is used to get the threshold to reduce the noise in the baseline.

$$P_{th} = \begin{cases} P_1 & R_1 > 1.5(R_2 + R_3) \ (R_1 = \mathbf{max}(r_i),\ R_2 = \text{mean}(r_i) \\ 0 & R_1 \le 1.5(R_2 + R_3) \ R_3 = \text{std}(r_i), \quad i = 1, 2, 3 \ldots m) \end{cases} \quad (11.2)$$

where, R_1 is the first max value of the sequence r_i, R_2 is the mean value of the sequence, R_3 is the standard deviation of the sequence, Pth is the power threshold for noise reduction in the baseline. The power of the data beneath the Pth is set to zero.

The mainly noise of Fig. 11.2b was seemed to be optical noise and fabrication error. It could not use a mean value smoothing method directly to reduce the mutations simply and efficiently. The method combining median filter and a mean value smoothing method with a short window length could reduce the mutations and smooth the curve efficiently. It was important to select a proper window length of the filters. Using a long window length, the method of mean value smoothing filter could easy distort the shape of the spectrums. A average method with proper window length would deal with the signal with little distortion.

$$y_k = \begin{cases} x_k & \max(W[x_k]) > x_k > \min(W[x_k]) \\ \mathbf{med}\,(W[x_k]) & x_k > \max(W[x_k]), x_k < \min(W[x_k]) \end{cases} \quad (11.3)$$

where, y_k is the data after filtering, x_k is the power of each wavelength in the spectrum, $W[x_k]$ is an window operation to the data, $\text{med}(W[x_k])$ is the median data of the data array of $W[x_k]$, $\min(W[x_k])$ is the minimum data of the data array of $W[x_k]$, $\max(W[x_k])$, is the maximum data of the data array of $W[x_k]$.

Compared Figs. 11.4 and 11.2b, the curve was smooth, and it was easy to find the peak position after the noise reduction process. And it was a base for the next procedure. These procedures were simple, efficient and easy. It could also archive in real-time.

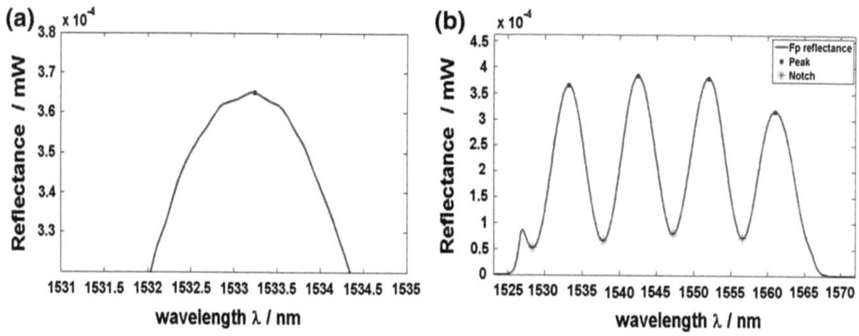

Fig. 11.5 The search results

11.3 Peaks and Notches Search

The procedures before were efficiently to reduce the noise and smooth the curve, and then it was base of the next program. In the paper, the first order derivation (the forward difference) was used. And the equation was as follow:

$$\Delta y_k = y_{k+1} - y_k, k = 1, 2, 3 \ldots l \quad (11.4)$$

where, Δy_k is the first order difference result of the signals. When Δy_k equal to zero, then the extreme point is judged by the signs of Δy_{k-1} and Δy_{k+1}, if the signs was difference and $\Delta y_{k-1} > \Delta y_{k+1}$ then the point is the peak, while $\Delta y_{k-1} < \Delta y_{k+1}$ then the point is the notch. The method was used to search the peak of Fig. 11.4 and the result was shown in Fig. 11.5a.

The derivation is sensitive to the noise in the Fig. 11.2. So the procedures before are the base for the peaks and notches search. From the theory it could find that the result of the peaks and notches search method used in fiber Fabry-Perot interferometer was good as shown in Fig. 11.5b. And it could find that all the peaks and notches were searched.

Two peaks in the spectrum of the Fabry-Perot interferometer sensor was to detected by the method as shown in Fig. 11.6, and it could find that the fitting line of the search result was quite linearity and fit to the sensing property of the sensor. And the spectrum signals was added noise artificially to test the method, and the signal noise ratio (SNR) was 40 dB, then search for 25 times and the mean square error was 0.055 nm because the resolution of spectrum was about 0.04 nm. So the deviation of the peaks was about 1–2 point of data.

11.4 Conclusion

A simple method proposed in the paper was quick, efficient and easy. And it could efficiently search the peaks and notches of the fiber Fabry-Perot interferometer sensor. From Figs. 11.5 and 11.6, the search result of peaks and notches of the

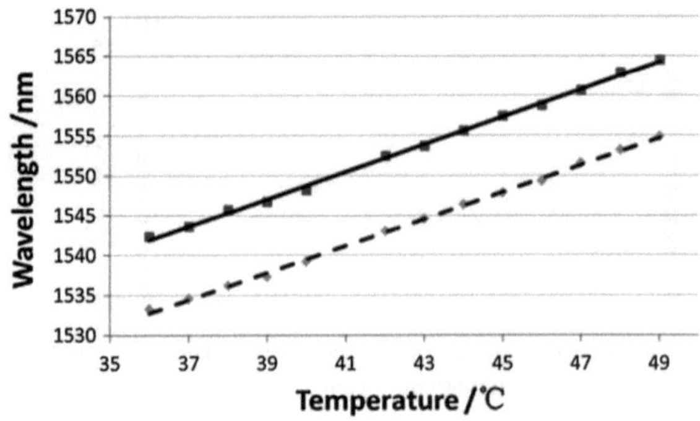

Fig. 11.6 The search result of two peaks in the spectrum of the senor

actual signals from the sensor showed that the method was feasibility. And the mean square error of the search results was about 0.055 nm for 25 times when the actual data was added noise to bring down SNR to 40 dB. And it could also be extended to the other optical spectrum signals.

Acknowledgments This work was supported in part by the National Natural Science Foundation of China under Grant No. 61203206 and 61273059, and the Fundamental Research Funds for the Central Universities under Grant No. 110604009 and 110804003, and IAPI Fundamental Research Funds under Grant 2013ZCX02-05.

References

1. Baek SJ, Park A, Kim J, Shen AG, Hu JM (2009) A simple background elimination method for Raman spectra. Chemometr Intell Lab Syst 98:24–30
2. Baek SJ, Park A, Shen AG, Hu JM (2011) A background elimination method based on linear programming for Raman spectra. J Raman Spectrosc 42:1987–1993
3. Yongqian L, Yang X, Guozhen Y (2010) Comparison of peak searching algorithms for wavelength demodulation in fiber Bragg grating sensors. In: Proceedings of 2nd international conference on information engineering and computer science (ICIECS) 2010, pp 1–4
4. Korun M, Vidmar T, Vodenik B (2008) Improving the reliability of peak-evaluation results in gamma-ray spectrometry. Accred Qual Assur 13:531–535
5. Likar A, Vidmar T (2003) A peak-search method based on spectrum convolution. J Phys D-Appl Phys 36:1903–1909
6. Li Q, Liao H, Liu X (2007) Novel peak-search algorithms of Raman spectrum. In: Proceedings of 8th international conference on electronic measurement and instruments (ICEMI '07). pp 2392–2395

Chapter 12
Magnetostriction Induced Fiberoptic Metal Profile Detector

Wei-Chih Wang, Benjamin Estroff and Wei-Shu Hua

Abstract This paper describes the design and construction of a magnetostrictive composite-fiber-optic Mach-Zehnder interferometer capable of magnetic field and metal profile measurement. Unlike previous metal detectors, the sensor makes use of the magnetostriction effect on a fiber-optic interferometer to detect metallic objects. The metal detector overcomes many difficulties existing in conventional metal detectors. Aside from offering relatively high sensitivity (sensitivity of about 70.7×10^{-3} rad/gauss), the optical detection provides resistance to RF interference which is common in typical electromagnetic type metal detectors. The magnetostrictive sensor is also relatively compact and able to achieve a minimum spatial resolution of 1 cm^2 for metal profile detection.

Keywords Fiberoptic sensor • Metal detector • Metal profile detector • Magnetostriction • Magnetostrictive polymer

12.1 Introduction

Metal detectors are instruments that use electromagnetic induction to detect metal objects. The first metal detector was invented by Alexander Graham Bell in 1881 and was used for detecting bullets when President James Garfield lay dying due

W.-C. Wang (✉) · B. Estroff · W.-S. Hua
Department of Mechanical Engineering, University of Washington, Seattle, WA, USA
e-mail: abong@u.washington.edu

W.-C. Wang
Department of Electrical Engineering, University of Washington, Seattle, WA, USA

W.-C. Wang
Medical Device Innovation Center, National Cheng Kung University, Tainan, Taiwan

R. Tutsch et al. (eds.), *Progress in Optomechatronic Technologies*,
Lecture Notes in Electrical Engineering 306, DOI: 10.1007/978-3-319-05711-8_12,

to the shot by assassin. In 1925, Gerhard Fischer made improvements to Bell's initial concept and invented a portable metal detector which was used to detect small pieces of buried metal. Fischer's metal detector became commercially available in 1931. Metal detection technology improved over several decades, creating more functional and effective devices. During World War II, the metal detector was extensively used in landmine detection. Currently, metal detector systems are prevalent in aspects of everyday life throughout the world. They are primarily utilized in the fields of defense and security, where they may be used to detect landmines or suspicious objects at security checkpoints. Additionally, the metal detector is used by geophysicists and archaeologists, and provides millions with an opportunity to discover hidden treasures, such as buried gold and silver. They are commonplace in libraries, airport security, historical sites, prisons, stores, armed forces, shops and government buildings.

Currently, metal detectors can be categorized into three basic technologies: very low frequency (VLF), pulse induction (PI), and beat-frequency oscillation (BFO) [1, 2]. These conventional metal detector systems share several disadvantages; they are unable to detect the profile of a metal object, are relatively bulky in size, and are not resistant to RF interference. In this paper, we describe the development of a novel compact metal detector system that overcomes these weaknesses. As we have previously published [3], our metal detector system is an integrated fiber-optic Mach-Zehnder interferometer with novel magnetostrictive material. The main concept of our metal detector is the use of a simple DC magnetic field scheme with the magnetostrictive material as the sensing device. The overall main structure consists of fiber-optic Mach-Zehnder interferometers. Thus, our sensor can be regarded as a novel metal detector system.

Our detector uses the change in length of a magnetostrictive material (WC-WSJOSHNICK-1) coating an optical fiber to alter the fiber's optical path length in the presence of a magnetic field. Thus, metal can be detected by monitoring the strain-induced optical path length change from the magnetostriction-induced photoelastic effect [4, 5]. Previous studies of these magnetostrictive material-coated fibers prove that this concept is not only successful for metal detection [6], but also show that our metal detector can identify the general geometrical shape of a rod, rectangular bar, wrench and steel plate with two holes [7, 8]. In this study, we report the latest results of the magnetostriction applications in a C clamp metal profile sensing.

12.2 Theory

Magnetostrictive materials are characterized by a longitudinal change in length due to an applied magnetic field, a phenomenon discovered by Joule [4, 5]. The recent increased interest in magnetostrictive sensor technology is a result of improved performance in magnetostrictive materials and new-found applications, including motion sensing, force and torque measurement, magnetic field detection, etc. In this paper, magnetostrictive material is used to fabricate a novel metal-detecting sensor. Our metal detector system consists of an integrated fiber-optic Mach-Zehnder interferometer with a sensing arm coated with magnetostrictive material. When the sensor experiences a change in the magnetic field of a metal object, it alters the optical path length of the interferometer through the magnetostriction-induced photoelastic effect.

Measurement of the intensity of the resulting strain on the magnetostrictive material is based on the magnetic field induced phase shift in the interferometer given by [9],

$$\Delta\phi = \frac{2\pi nL}{\lambda}\left(\frac{\Delta L}{L} + \frac{\Delta n}{n}\right) = \frac{2\pi nL}{\Delta}\left\{\varepsilon_3 - \frac{n^2}{2}\left[(p_{11} + p_{12})\varepsilon_1 + p_{12}\varepsilon_3\right]\right\} \quad (12.1)$$

In Eq. (12.1), longitudinal strain $\Delta L/L = \varepsilon_3$, ε_1 is transverse diagonal strain index of refraction based on the phase retardation of the propagating light in the fiber caused by the strain which is induced by the B field, using the form of a photoelastic tensor p_{ik} is given as $\Delta n = -n^3/2\left[(p_{11} + p_{12})\varepsilon_1 + p_{12}\varepsilon_3\right]$. The fiber is assumed to be an isotropic medium with the light polarizing along the transverse direction. Using the data for fused quartz, $p_{11} = 0.12$, $p_{12} = 0.27$, $n = 1.46$ and $\varepsilon_1 = -\varepsilon_3/2$ (assuming zero net-volume change) [9], Eq. (12.1) reduces to:

$$\Delta\phi = \frac{2\pi nL}{\lambda} 0.92\,\varepsilon_3 \quad (12.2)$$

where the photoelastic effect due to ε_1 and ε_3 nearly cancel such that the main contribution is due to physical elongation of the fiber $\varepsilon_3 = kH^2$, where k is a constant for the ferromagnetic material used. For this study, ε_3 will be determined empirically.

12.3 Sensor Configuration

The fiber-optic sensor is coated with magnetostrictive material and embedded in composite resin. Numerous fiber-optic magnetic field sensors have been developed. Yariv and Winsor [9] proposed a now common configuration which uses a magnetostrictive film coating an optic fiber. This sensor is in the form of two Mach-Zehnder fiber-optic interferometers immersed in a magnetic field. The magnetic field leads to the magnetostrictive film to deform, straining the optic fiber. This causes a change in the length of the optical path. An interferometer is used to measure the phase changes. Our metal detector research concentrates on this technique; please refer to our previous experiments [7, 8] for greater detail. In this paper, the metal detector utilizes a single mode fiber with the ferromagnetic polymer coating applied directly onto the cladding of the sensing arm. Measurement of the intensity of the resulting strain on the magnetostrictive material is based on the magnetic field induced phase shift in the interferometer given by [9]. The structure of magnetostrictive metal sensor is shown in Fig. 12.1a. We used a different layout

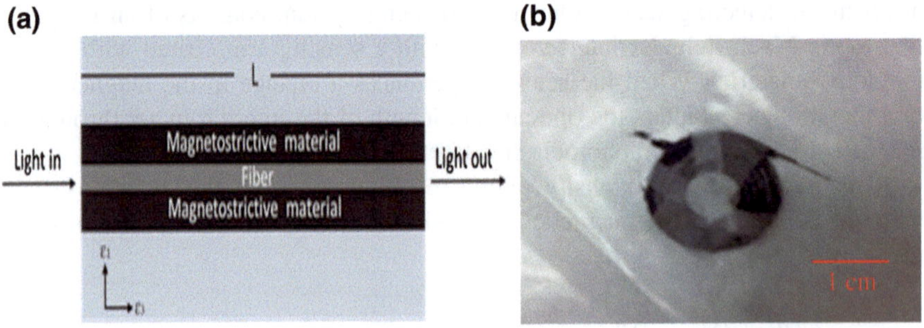

Fig. 12.1 Configuration of metal sensor: **a** the structure of magnetostrictive metal sensor, **b** the photo of magnetostrictive metal sensor (WC-WSJOSHNICK-1)

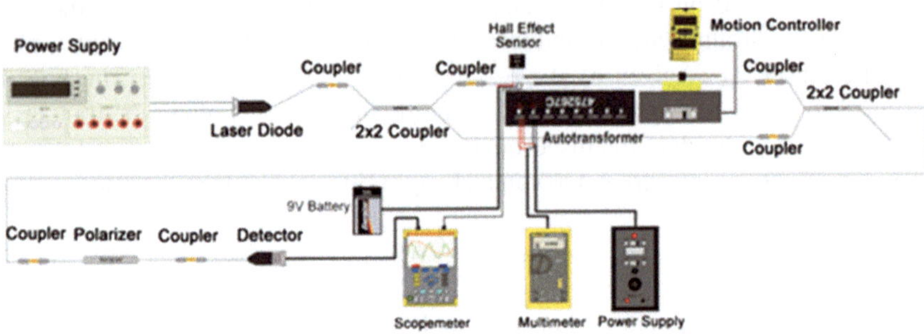

Fig. 12.2 Diagram of polymeric fiber-optic magnetostrictive metal detector system

and geometry than previously reported [8]. In this case, the sensing fiber is bent into a circle shape in order to reduce the detector size (30 cm long and 1.5 mm thick, see Fig. 12.1b, and the magnetostrictive material is manipulated during fabrication to increase its sensitivity.

12.4 Experimental Setup

The overall metal detector system is based on a Mach-Zehnder fiberoptic interferometer, the sensing element for detecting changes in magnetic fields consisting of magnetostrictive material coating an optical fiber of an interferometer arm. When a change in the magnetic field occurs it causes longitudinal dimension changes in the magnetostrictive material, inducing a strain in the optical fiber. This strain causes a phase shift in the optically propagating beam in the fiber, which is detectable by interferometry. The general experimental setup for the magnetostrictive metal detector system is shown in Fig. 12.2. For greater detail in the components used and experiment setup, please refer to our previous publications [8].

In this paper, the metal object (a C-clamp) is mounted on a wooden stage with guiding rails (Fig. 12.3). The clamp was attached on the underside of the cart,

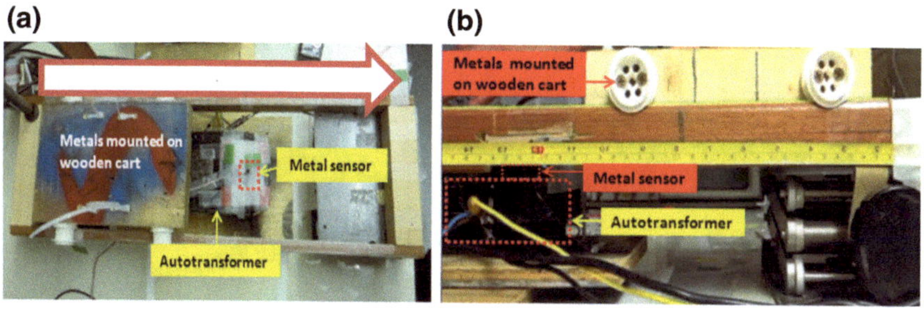

Fig. 12.3 The Configuration of wooden track stage. **a** Overhead view of the stage with the metal sensor. **b** Side view of wooden track stage

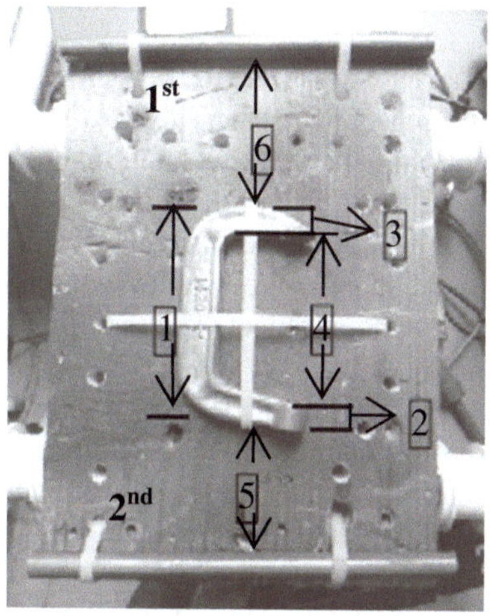

Fig. 12.4 Underside of cart (see Table 12.1)

Table 12.1 Sample Dimensions

Investigated samples	Position	Width (cm)
Rod	1st rod	0.9
	2nd rod	0.9
Clamp	Clamp's arm (1)	8.5
	Clamp south (2)	1.3
	Clamp north (3)	1.3
Gap	Distance middle (4)	6.7
	Distance south (5)	5
	Distance north (6)	6.5

parallel to the metal sensor (Fig. 12.4). Similar to the previous metal detector setup [8], a constant DC magnetic field is applied to the sensor, with a Hall Effect sensor positioned directly above for a reference signal (~1 cm above the auto-transformer). When a metal object is directly above the transformer, a localized magnetic disturbance is created. This disturbance results in a phase shift that is observed in the interferometer output. Table 12.1 provides details on the actual size and position of the metal object. The space between the metal clamp and

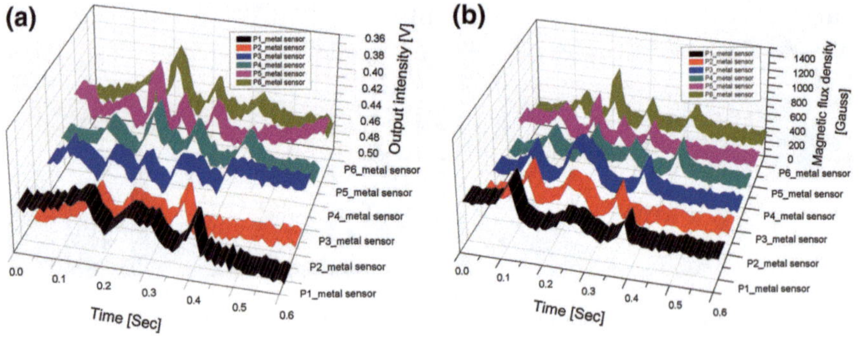

Fig. 12.5 **a** Plots of output intensity from magnetostrictive metal sensor (WC-WSJOSHNICK-1) with C clamp. **b** Plots of magnetic flux density from Hall Effect sensor with C clamp

the sensor is 1 cm. The cart was moved back and forth by hand, striving for a constant speed. For profile mapping, the cart with the clamp was shifted in 1 cm increments from left to right. In repeated experiments, this design was not only able to detect the presence of metal objects, but was also capable of mapping metal shapes. We use the results of these trials to determine the shape and size of the metal object. The resulting signals from both the metal sensors are shown in Fig. 12.5.

For the metal detector testing, a DC magnetic field was fixed at 500 gauss (corresponding input DC voltage of 4 V from the autotransformer). The output sensor signal is obtained and compared with Hall Effect sensor (Fig. 12.5). As we know, the interferometer signal is sine wave; the result of P3 (Fig. 12.5a) shows opposite tread line due to the signal was captured at the area of descending sine wave. Due to the limitations of sensor size and sensing area, we only could detect one trial each time. Another issue is the starting detection time; as we could not control the speed of the movement wooden cart by hand, and the starting time is relative to the velocity of the cart, each of the six trials began at different points. We compared six metal detection trials from our metal detector (WC-WSJOSHNICK-1) and the Hall Effect sensor. The results (Fig. 12.5a, b) clearly show peaks in the signal when the metal is over the sensor. In addition, the peaks appear as different shapes due to the different positions of the detected metals, and show clear matches between the sensors. Peaks with different width and amplitude correspond to different size and thickness of the metal. The general outline of the metal can be seen from the 3D plot. Compared to the results of the Hall Effect sensor, we can summarize that the output intensity from the magnetostrictive metal sensor (WC-WSJOSHNICK-1) shows a sawtooth curve, which can also be observed in Fig. 12.5a, when the metal's surface bumps are detected. Therefore, the metal sensor successfully shows the profile of the C clamp. In addition, Fig. 12.5a, b demonstrate that the metal sensor is relatively more sensitive to magnetic disturbances than Hall Effect sensor. Based on the preliminary results, this sensor is also capable of determining the

Table 12.2 The calculation result of investigated samples

Investigated samples	P1 (cm)	P2 (cm)	P3 (cm)	P4 (cm)	P5 (cm)	P6 (cm)	Actual size (cm)	Average error (%)
1st rod	0.86	1	1	1	1	1.24	0.9	14.44
2nd rod	0.98	1	1	1	1	1	0.9	10.74
Clamp's arm (1)	8.59	9.16	N\A	N\A	N\A	N\A	8.5	4.41
Clamp south (2)	N\A	N\A	1.6	1	1.24	1.24	1.3	13.84
Clamp north (3)	N\A	N\A	1.24	1.6	1.33	1.04	1.3	12.5
Distance middle (4)	N\A	N\A	6.72	6.48	7.2	7.2	6.7	4.62
Distance south (5)	5.25	4.86	4.8	5.04	5.04	4.86	5	2.7
Distance north (6)	6.33	6.72	6.42	7.2	7.2	6.72	6.5	5.35
Distance between rods	19.74	20.33	21.2	21.4	20.25	20.53	20.4	2.29

metal's irregular shape. However, our results show that our metal sensor (2.3 cm diameter) has superior resolution compared to the commercial Hall Effect sensor (0.25 cm^2). We also notice that based on the signal-to-noise (S/N) ratio, our sensor is more sensitive than the Hall Effect sensor. Again, this ratio depends on the S/N ratio as well as the permeability of the metal and the DC magnetic field present in the sensor. In future designs, the sensing performance will be improved by reducing the sensor loop diameter. In addition, the method of moving metal objects will be improved to avoid vibration interference, utilizing an automated stage to move the objects at a constant speed and stabilize the baseline intensity. The fluctuation in both signals may be caused by temperature perturbations and slight vibration.

Our previous publication [8] details how the fiber sensor successfully detects a steel wrench and steel plate. The results from [8] show that the magnetostrictive metal sensor can roughly distinguish the metal's profile. In this paper, we also discuss the broadened application of the sensor: the ability to create a 2D outline of an object. The metal object used for testing was an irregularly shaped metal clamp (Fig. 12.4). In publication [8], it was proved that the size of the metal object could be calculated based on the period of time it was detected and the velocity of the object being detected. Because our sensor was 2.3 cm in diameter and the metal object was approximately 5 cm in width, the object was scanned in six segments. Using the resultant output signals and velocities from each segment, we could calculate the size of the metal object; the results are shown in Table 12.2. The values calculated were very close to the actual dimensions of the objects. The average percent errors between the calculated distances and the actual distances ranged from 2.29 to 14.44 %. According to the calculated results (Table 12.2), we could plot the metal object's outline using AutoCAD (Fig. 12.6a). Based on Fig. 12.6a, it shows that the metal sensor (WC-WSJOSHNICK-1) is capable of determining the metal's irregular shape. Due to the limitation of sensor size and sensor detection area, we could only detect the rough shape of the metal.

Improvement is underway to remove the need to keep the constant scanning speed and reduce the sensing area. It will be published in the near future.

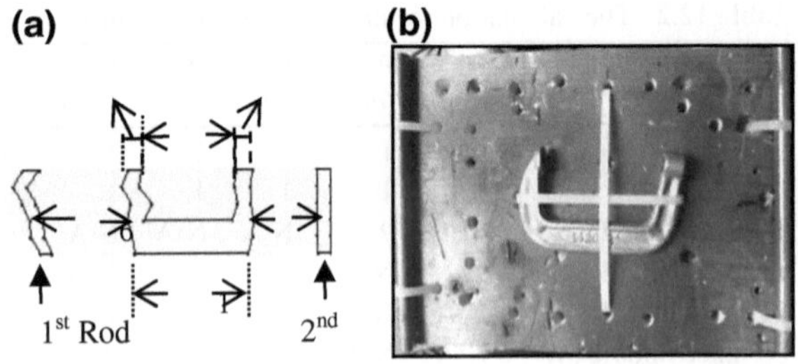

Fig. 12.6 Plots of output mapping result from magnetostrictive metal sensor with C clamp. **a** Mapping metal's outline. **b** Actual investigated metals

12.5 Conclusion

In this paper, we present a polymeric magnetostrictive fiber-optic metal detector system using a magneostrictive material (WC-WSJOSHNICK-1). The results from the fiber-optic system show that the magneostrictive materials work well as a metal sensor. In addition, the sensor can roughly identify irregular metal shapes, such as a steel rod, steel plate, steel wrench and now a C clamp. The outline of metal objects can also be constructed based on the output signal. The experimental results show the sensor has a minimum detected area of <0.9 cm and a sensitivity of 70.7×10^{-3} rad/gauss.

Acknowledgement This work is supported in part by NIH grant R01 EB007636-01A4.

References

1. Sharawi MS, Sharawi MI (2007) Design and implement of a low cost VLF metal detector with metal-type discrimination capabilities. In: Proceedings of IEEE international conference on signal processing and communications, Dubai, United Arab Emirates, 24–27 Nov 2007
2. Payne GC (1985) Metal detector system for identifying targets in mineralized ground. US Patent 4,507,612, Mar 26 1985
3. Hua WS, Hooks J, Wu WJ, Wang WC (2010) Development of a novel polymeric fiber-optic magnetostrictive metal detector. SPIE health monitoring of structural and biological systems, San Diego, CA, 7–10 Apr 2010, SSN10-82
4. Joule JP (1842) On the new class of magnetic force. Ann Electr Magn Chem 8:219–224
5. Joule JP (1847) On the effect of magnetism upon the dimension of iron and steel bars. Phil Mag 30:76–92
6. Hua WS, Hooks J, Wu WJ, Wang WC (2010) Development of a polymer based fiberoptic magnetostrictive metal detector system. In: Proceedings of international symposium on optomechatronic technologies (ISOT 2010), Toronto, Ontario, Canada, 25–27 Oct 2010
7. Hua WS, Wang WC (2011) Development of a ferromagnetic polymeric metal detector system. SPIE health monitoring of structural and biological systems, San Diego, CA, Mar 6–10 2011
8. Hua WS, Hooks J, Wu WJ, Wang WC (2011) A polymeric fiberoptic metal sensor capable of profile detection (ICAST 2011), 10–12 Oct 2011 at Imperial Hotel, Corfu, Greece
9. Yariv A, Windsor H (1980) Proposal for detection of magnetic field through magnetostrictive perturbation of optical fibers. Opt Lett 5:87

Chapter 13
Liquid Viscosity Sensing Using Nonlinear Higher Order Harmonic Excitation in Fiberoptic Vibrating Sensor

Wei-Chih Wang and Per G. Reinhall

Abstract In practice, the vibration amplitudes are always kept sufficiently small in order to avoid nonlinear vibration effects. In this paper, however, nonlinear vibration is intentionally excited to improve the sensitivity of viscosity measurements, in particular nonlinear higher order harmonic excitation. The results show that a nonlinear effect drastically improves sensitivity to viscous damping. Theory and experimental results are presented. Three different experiments were conducted: (1) observe the frequency responses of the third harmonic in a linear system; (2) excite the system into a nonlinear region using different driving (~2 kHz) and response frequencies (~6 kHz); (3) excite the system into a nonlinear region using the same frequency for both driving and response (~6 kHz). For liquid viscosity measurement, the gain per unit of viscosity for the superharmonic case is highest among all cases at 3 dB/cp.

Keywords Nonlinear • Vibration • Viscosity measurement • Higher order harmonic excitation • Superharmonic excitation • Subharmonic excitation • Fiber optic sensor • Forward light scattering • Nonlinear system

13.1 Introduction

In this paper, viscosity measurement based on nonlinear vibration of the forward light scattering fiber resonator is presented. The procedure involves deriving viscosity from the nonlinear frequency response of the vibrating fiber viscometers by

W.-C. Wang (✉) · P. G. Reinhall
Department of Mechanical Engineering, University of Washington, Seattle, WA 98195, USA
e-mail: abong@u.washington.edu

W.-C. Wang
Department of Electrical Engineering, University of Washington, Seattle, WA 98195, USA

W.-C. Wang
Medical Device Innovation Center, National Cheng Kung University, Tainan, Taiwan, ROC

R. Tutsch et al. (eds.), *Progress in Optomechatronic Technologies*,
Lecture Notes in Electrical Engineering 306, DOI: 10.1007/978-3-319-05711-8_13,

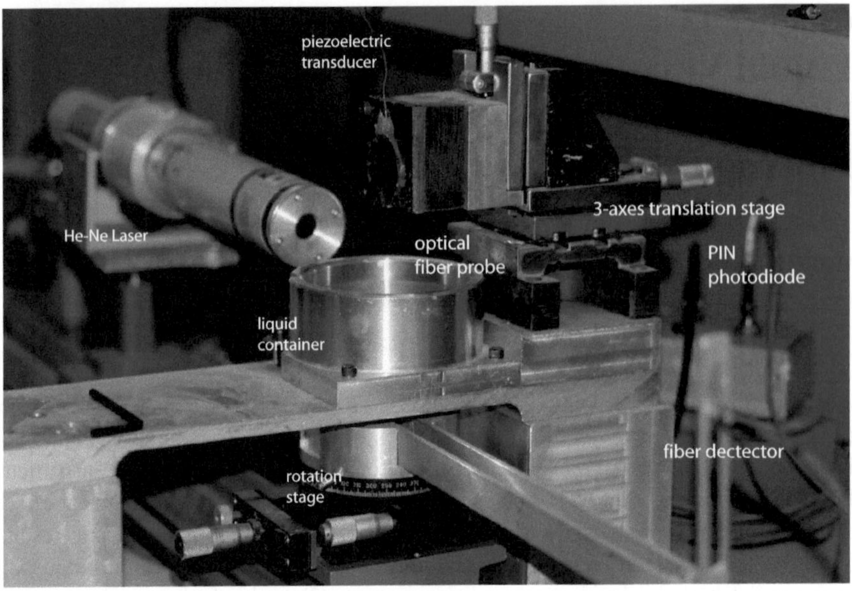

Fig. 13.1 Apparatus used to measure liquid viscosity

a large subharmonic force excitation. The set-up for the nonlinear vibration in the previous linear system [1–6] (Fig. 13.1) is slightly modified for the higher order harmonic response. In this case, the response is excited and picks up at the third harmonic frequency (f_3 ~ 6 kHz) of the vibrating fiber. The result of the nonlinear higher order harmonic response is carefully examined and compared with the linear primary and higher order harmonic response. The forced Duffing oscillator with quartic anharmonicity is still the model for the nonlinear vibration of the stretch fiber. Without further introducing the linear vibration equation, one could refer to the linear vibration theory presented in [1, 2]. Based on the theory, one will see that using a higher order response for viscosity measurement in either a linear or nonlinear configuration can result in a higher sensitivity than the previous linear main response described in [1, 2]. One will also find that the displacement response, aside from being a function of the fluid damping and the input amplitude, is a function of the response frequency.

13.2 Theory

For the case of nonlinear forced vibration, the resulting equation of motion appears as [7],

$$A_o\left(\omega_n^2-\omega^2\right)+\frac{3}{4}\varepsilon A_o^3=\sqrt{p^2-(\beta(\omega A_o))} \qquad (13.1)$$

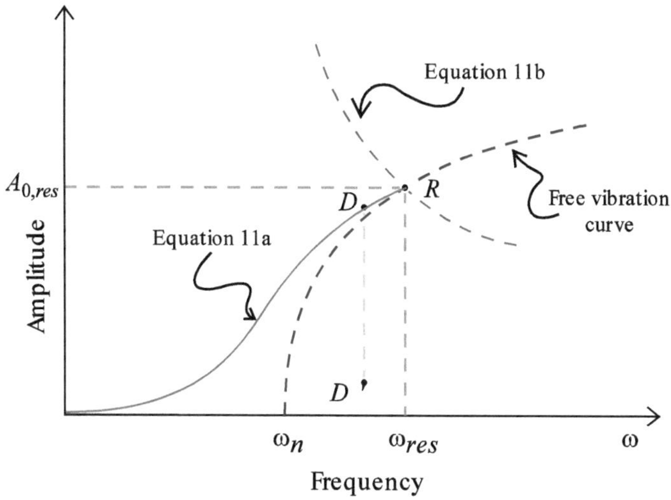

Fig. 13.2 Typical nonlinear response spectra in the case of a hardening stiffness

It is important to clarify the difference between the maximum amplitude and the resonance amplitude since the maximum amplitude no longer occurs at the resonance frequency. The resonance amplitude can be found by simultaneously setting the damping and forcing term of (11) to zero which gives the undamped amplitude equation $A_o\left(\omega_n^2 - \omega^2\right) + \frac{3}{4}\varepsilon A_o^3 = p$ [15]. This allows one to set the left side of (1) equal to zero and to solve for the resonance amplitude A_0 which is:

$$A_o = \frac{p}{\beta \omega_n} \tag{13.2}$$

Once again the resonance amplitude appears to be proportional to $1/\sqrt{\rho_l \mu}$ as in the linear model of the primary response case. Since the maximum point may not actually be attained (because of the hysteresis that occurs as indicated from point D to D' in Fig. 13.2), this rules out the possibility of true resonances. The damped vibration amplitude is therefore measured at the point where the jump occurs and also at the point where maximum amplitude occurs. Each technique is analyzed in the following discussion and different measuring schemes will be derived from each technique.

If the viscosity is measured based on the maximum amplitude, the maximum amplitude can be obtained by differentiation of (13.1) with respect to ω. The resulting maximum displacement appears as

$$A_o = \sqrt{\frac{p^2}{\left(\beta^2 \omega^2 + \frac{\beta^4}{4}\right)}} \tag{13.3}$$

Based on the above equation, it is clear why the maximum amplitude is more affected by damping β in the nonlinear system than in the linear system. Although

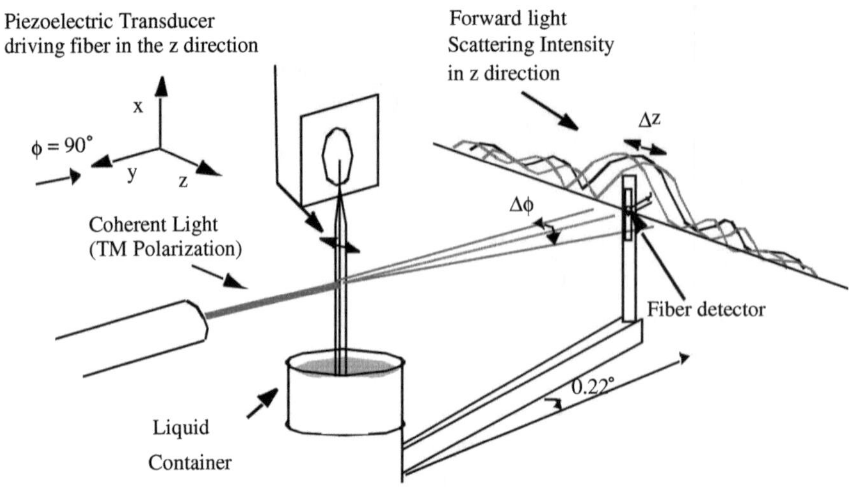

Fig. 13.3 Experiment setup for measuring liquid viscosity

the maximum vibration amplitude of both systems decreases linearly with increasing frequency, the additional constant $\beta^4/4$ in the nonlinear system will offset the amplitude even more. This is especially true in the case of soft structures vibrating in a highly viscous fluid. This effect was verified by experimentation.

13.3 Experiment Setup

The experimental set-up is repeated from the earlier forward light scattering technique described in [1, 2, 4]. A diagram of the experimental system is shown in Fig. 13.3. The system consists of a 10 mW He-Ne linearly polarized laser and a single mode fiber (2.75 μm diameter core and 125 μm diameter cladding). One end of the fiber is anchored to the bottom of a reservoir containing a liquid for testing, and the other end is attached to a piezoelectric transducer. The light from the laser is aimed normal to the single mode fiber and results in a light pattern scattering from the fiber. The displacement of the fiber is measured by the motion of the light intensity variations created by forward light scattering from the fiber. A small light detector (PIN diode, pigtailed to a photodetector) is placed where a strong change of brightness occurs in the scattering pattern. A linear relationship has been found experimentally between a relatively small displacement (<1 mm) and the corresponding intensity variation from the scattering pattern. This provides a simple method for measuring the vibration amplitude of the fiber. The viscosity of the fluid is deduced from the frequency response of the vibrating fiber. The top end of the optical fiber is vibrated by a piezoelectric transducer. The light intensity from the photodectector is collected by using an automated data acquisition system. The frequency response of the fiber is generated by a sinusoidal input sweep, allowing the viscosity of the fluid to be estimated using the peak amplitude of the frequency response.

13.4 Results and Discussion

The techniques used in this experiment were: (1) observe the frequency responses of the third harmonic in a linear system (Fig. 13.4a); (2) excite the system into a nonlinear region using different driving (~2 kHz) and response frequencies (~6 kHz) (Fig. 13.4b); (3) excite the system into a nonlinear region using the same frequency for both driving and response (~6 kHz) (Fig. 13.4c). The overall signals are less stable where noise levels fluctuated around 0.4 dB. The frequency drift is also high ($\pm$10 Hz) partly because the frequency resolution is lower at this frequency (2 Hz).

Based on the results, we found that the vibration amplitude in both linear and nonlinear primary resonance of higher order harmonics were even more susceptible to damping. For the nonlinear third harmonic response, a 5 dB increase or 50 % improvement was observed in the air to 40 % sucrose concentration measurement (Fig. 13.5). For the third harmonic driven at the first harmonic frequency, the improvement was 120 %. The most improved was in the linear third harmonic response, where the increase was 178 %. For the liquid viscosity measurement between 1 and 6 cp, both linear and nonlinear third harmonic responses showed a 3–11 dB increase compared with the linear first harmonic response. Among them, the third harmonic driven at the first harmonic frequency shows the most improvement among all the higher order harmonic cases. For liquid viscosity measurement, the gain per unit of viscosity for the superharmonic case is highest among all cases at 3 dB/cp.

When the normalized intensity is converted to vibration amplitude, we found that the linear third harmonic resonance has the lowest magnitude among the four we compared (Fig. 13.5a). This is expected since it is harder to generate high vibration amplitudes at higher order resonances. It is also due to the fact that the same input was given for all the linear system tests. Therefore it is expected that the third harmonic will have lower peak vibration amplitude than the first harmonic. Both the nonlinear and superharmonic cases exhibited larger vibration amplitudes (two orders higher) than the linear cases. This make sense because the displacement input for the nonlinear and superharmonic cases is about 150 times higher than the input for the linear first harmonic response. For the superharmonic and nonlinear third harmonic resonances, the peak vibration amplitude versus $1/\sqrt{\rho_l \mu}$ are nonlinear. Both curves also have two transitions points, one occurring around 10 % sucrose concentration, and the other at 30 % concentration.

From the nonlinear third harmonic response, we observed an interesting transition of the system when it gradually changed from nonlinear to linear due to the increased damping. This is seen by the disappearance of the hysteresis in the response curve after the concentration fell below 10 % (see Fig. 13.4c). Although the curve still appears nonlinear, the effect is probably reduced by damping. This gradual change from nonlinear response to linear response due to damping is also observed in the superharmonic excitation (Fig. 13.4b) where the jumps disappeared after water was used as the solution.

The experimental results confirm that the higher order harmonic responses are more susceptible to damping. Based on Eq. (13.3), we know that excitation

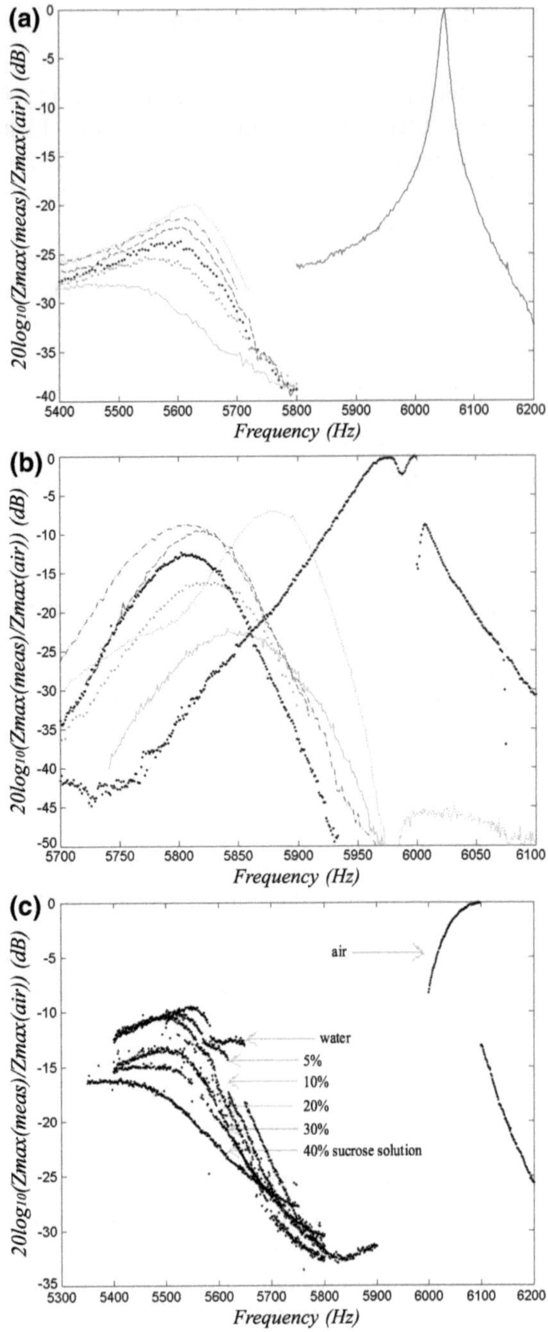

Fig. 13.4 Viscosity experiment: **a** linear primary resonances driving near its third harmonic frequency (fobserved ~ 5,800 Hz). **b** Superharmonic excitation observed at 3rd harmonic resonance (fexcitation ~ 2,000 Hz, fobserved ~ 5,800 Hz). **c** Nonlinear primary resonances (fobserved ~ 5,800 Hz). Fluid samples are air (*dark solid*), deionized water (*dotted*), 5 % (*dashdotted*), 10 % (*dashed*), 20 % (*point*), 30 % (*light point*), 40 % (*light solid*) sucrose solutions. Due to hysteresis, Plot (**c**) curves were plotted with points and legends are replaced with labels as shown in the figure

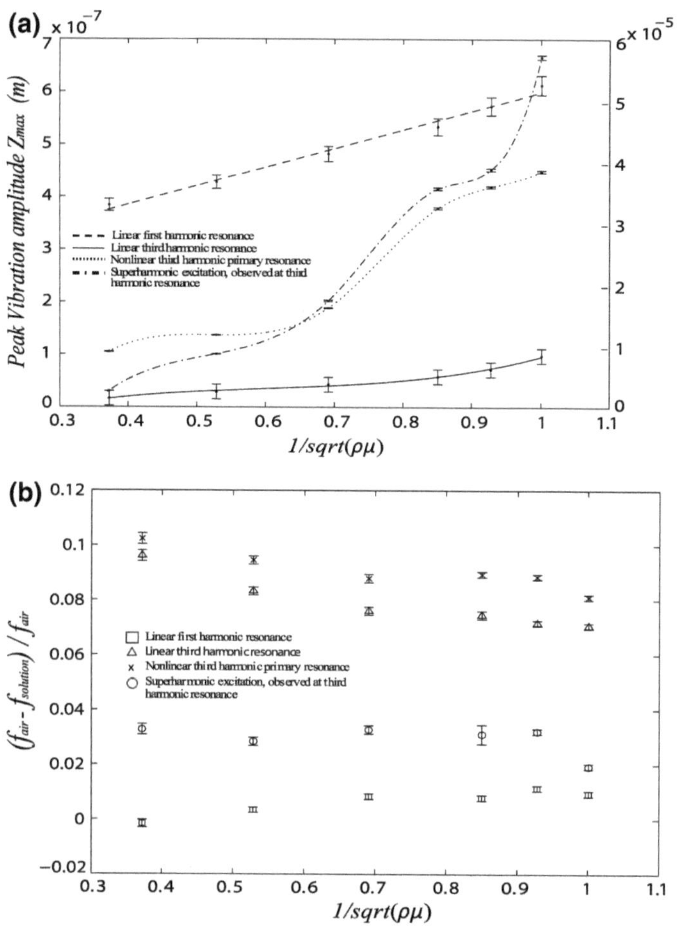

Fig. 13.5 **a** Peak vibration amplitude versus 1/sqrt(ρμ). **b** Frequency shift versus 1/sqrt(ρμ). The plot in (**a**) consists of two vertical axes: axis on the left is for the linear first harmonic resonance (*dash*) and third harmonic resonance (*solid*), axis on the right is for nonlinear third harmonic primary resonance (*dotted*) and superharmonic excitation observed at third harmonic (*dashdotted*). Errors are smaller than the size of symbols used for data points in the plots. Errors are smaller than the size of symbols used for data points in the plots. Air measurement is not included in the plot

of higher order harmonics does indeed increase the damping in both linear and nonlinear systems. Since the constant β^4/i in the nonlinear system is relatively small compared to the $\beta^2\omega^2$, the linear response equation can be substituted for the nonlinear response. Based on $A_o = 2p/(\omega\beta)$, one can expect that the third harmonic resonance amplitude will undergo a larger change than the first harmonic resonance because the frequency is three times larger. As expected, we observed an increase in damping for both the linear and nonlinear third harmonic response (Fig. 13.5a). Based on the curve, the amplitude of the vibration appears higher for the lower frequencies. This confirmed the findings where the linear main response is riding above the primary resonance of the third harmonic. We also found the rate of change in vibration amplitude versus $1/\sqrt{\rho_l \mu}$ is very close to the nonlinear primary resonance and the linear third harmonic main response. This justifies the assumption that the $\beta^4/4$ term in the nonlinear system cannot contribute any significant amount of damping to the nonlinear system.

The change in the direction of the frequency shift due to increases in viscosity did not appear in either the nonlinear or the linear third harmonic response (Fig. 13.5b). This was very different from the results we got from the linear first harmonic and the superharmonic excitation where a change in direction of the resonant frequency occurred, actually increasing with higher sucrose concentrations. The patterns of the frequency shift for both third harmonic responses are very similar. However, the magnitude of the frequency shift is slightly larger for the nonlinear case. This is expected since the resonance frequency (defined as the frequency with the maximum amplitude) for a hardening nonlinear system occurs at higher frequencies than a linear case. Again, the pattern of frequency shift patterns observed in the linear and superharmonic resonances also appears in the superharmonic excitation of the third harmonic (Fig. 13.4b).

13.5 Conclusion

In this paper we presented our funding in viscosity sensing using linear and nonlinear higher order harmonic excitation. Based on the results, we found that the vibration amplitude in both linear and nonlinear primary resonance of higher order harmonics are more susceptible to damping. For the nonlinear third harmonic response, a 5 dB increase or 50 % improvement was observed in the air to 40 % sucrose concentration measurement. For the third harmonic driven at the first harmonic frequency, the improvement was 120 %. The most improved was in the linear third harmonic response, where the increase was 178 %. For the liquid viscosity measurement between 1 and 6 cp, both linear and nonlinear third harmonic responses showed a 3–11 dB increase compared with the linear first harmonic response. Among them, the third harmonic driven at the first harmonic frequency shows the most improvement among all the higher order harmonic cases. For liquid viscosity measurement, the gain per unit of viscosity for the superharmonic case is highest among all cases at 3 dB/cp.

References

1. Wang W-C, Reinhall P, Yee S (1999) Fluid viscosity measurement using forward light scattering. Meas Sci Technol 10:316–322
2. Wang W-C, Yee S, Reinhall P (1995) Optical viscosity sensor using forward light scattering. Sens Actuators, B 24–25:753–755

3. Wang W-C (1996) A study of fluid viscosity and flow measurement using fiber-optic transducer. Ph.D. Thesis, University of Washington, U.S.A
4. Fedorchenko AI, Stachiv I, Ho J, Wang A, Wang W-C (2008) On the forced vibration of the fiber partially immersed in fluid. Sens Actuators, A: Phys 147(2):498–503
5. Fedorchenko AI, Stachiv I, Wang AB, Wang W-C (2008) Fundamental frequencies of mechanical systems with *N*-piecewise constant. J Sound Vib 317:490–495
6. Fedorchenko AI, Stachiv I, Wang W-C (2013) Method of the viscosity measurement by means of the vibrating micro-/nanomechanical resonators. Flow Meas Instrum 32:84–89
7. Timoshenko S, Young DH, Weaver W Jr (1974) Vibration problems in engineering, 4th edn. Wiley, New York

Part IV
Optomechatronics Applications in Bio-Medical, Non-Human Surgical Assistants, Manufacturing and Robotics

Chapter 14
Simulation and Optimization of Nanoparticle Patterned Substrates for SERS Effect

Mei Liu, Yan Peng and Zhizheng Wu

Abstract In order to make the best of the surface enhancing behaviors of metal nanostructures for Raman Scattering, a tactful balance should be found between signal enhancement, the distribution uniformity of 'hot spots' and the reproducibility of nanostructure patterned substrates, which should generally be testified by simulation and experiment. This paper simulated and compared the Raman enhancements produced from a variety of nanoparticle covered SERS substrates with different sizes and spaces, and it was concluded that the distance between the nanoparticles plays a contradictory role on the enhancement factor and the uniformity of the 'hot spots', and so it should be selected with comprehensive consideration.

Keywords Raman scattering • Surface enhancement Raman scattering (SERS) • Nanoparticles • Substrates

M. Liu (✉) · Y. Peng · Z. Wu
School of Mechatronics Engineering and Automation, Shanghai University, Shanghai 200072, People's Republic of China
e-mail: mliu@shu.edu.cn

Y. Peng
e-mail: pengyan@shu.edu.cn

Z. Wu
e-mail: zhizhengwu@shu.edu.cn

M. Liu
State Key Laboratory of Transducer Technology, Shanghai Institute of Microsystem and Information Technology, Chinese Academy of Sciences, Shanghai 200050, People's Republic of China

R. Tutsch et al. (eds.), *Progress in Optomechatronic Technologies*,
Lecture Notes in Electrical Engineering 306, DOI: 10.1007/978-3-319-05711-8_14,

14.1 Introduction

Surface-enhanced Raman scattering (SERS) is playing an increasing important role in biomedical research, biophysics and biochemistry, including single molecule detection, non-intrusive study of reaction dynamics, and also identification of tiny amounts of biological molecules or dangerous chemical species [1–3].

The fundamental physics of SERS has been extensively studied and generally understood [4]. Basically, the excitation of surface plasmons (SPs) in metal nanostructures can generate sizable electromagnetic field enhancements due to large transient surface dipoles induced by plasmons. Based on this principle, various types of substrates patterned with fine geometric nanostructures have been designed to increase the enhancement factor, and high SERS enhancements have been observed at certain spots of a substrate, making the detection of analytes possible at extremely low concentrations [5–9].

To provide more insight and understanding of the Raman scattering enhancement observed in previous publications, we have performed electromagnetic calculations by a finite element method using the commercial COMSOL multiphysics software package. For simplicity, the system was modeled as solid objects next to each other, which usually results in underestimating of actual field enhancements since local imperfections can enhance local fields drastically.

14.2 Establishment of the Models

In this paper, two types of simplified gold nanoparticles (NPs) models were constructed, simulated and compared to optimize substrates for SERS. The Gaussian electromagnetic wave follows the Maxwell equations in the simulated area, and boundary conditions at the interface between the scattering NPs and the medium. The scattering boundary condition, adsorption boundary condition and perfect matching layer condition was used in the simulation process. For SERS, it is generally agreed the Raman intensity increases by a factor $|E|^4$ and the Raman enhancement factor could be evaluated as [10]:

$$G(r, \omega) = \left| \frac{E(r, \omega)}{E_{inc}(r, \omega)} \right|^4 \tag{14.1}$$

$$AverageG(\omega) = \left| \frac{AverageE(\omega)}{E_{inc}(\omega)} \right|^4 \tag{14.2}$$

where G and AverageG are two different factors to evaluate the maximum enhancement factor and the average enhancement factor in a fixed domain, $E(r, \omega)$

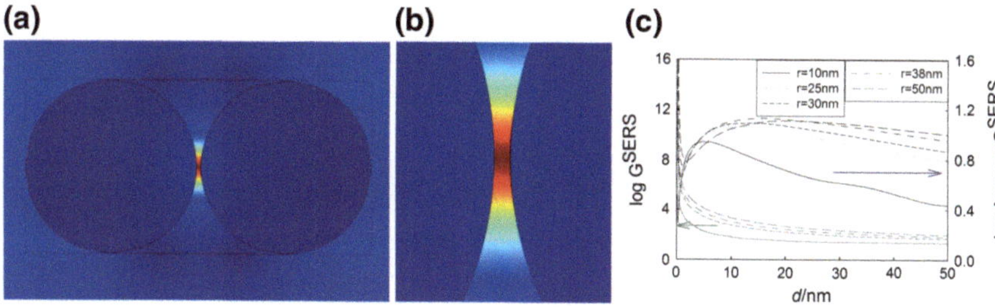

Fig. 14.1 The over-all electric field distribution between two NPs. **a** The electric field between two NPs. **b** The enlarged figure of the middle part of **a**. **c** The enhancement factor and average enhancement factor as a function of the distance d. *Arrows* in the figure indicates the respective axis each group of curve is corresponding to

is the overall electric field at location r, and $E_{inc}(r, \omega)$ is the electric field relating to the incident electric wave, AverageE(ω) is the average electric field strength in the fixed domain, and $E_{inc}(\omega)$ is the average electric field relating to the incident electric wave. The incident electromagnetic wave is polarized along the centerline of the NPs. The spot radius of the incident wave is equal to the laser wavelength, and the background electric field amplitude is 1 V/m.

14.2.1 Two Nanoparticles (NPs)

Here the electric field distribution between two NPs was simulated and obtained, shown in Fig. 14.1, where the radius *r* of the NPs is 10 nm, the distance *d* between the two NPs is 0.5 nm, and the wavelength λ of the electromagnetic wave is 785 nm.

The maximum enhancement factor and average enhancement factor between two NPs as a function of distance is shown in Fig. 14.1c, where λ is 785 nm. It can be seen that both enhancing factors decrease as the distances increases, with a big difference of 6 orders in enhancing factor as the distances changes from 0.7 to 10 nm. And the enhancing factors of smaller NPs tends to change more quickly, whose maximum AverageG reaches maximal at a smaller distance.

The calculated SERS enhancement at the hot spot is found to be able to reach 107–108, which may be sufficient for detection of a few molecules at the resonant frequency of the molecule. And from previous publications it is known that the nanostructure with curvature and dissymmetry is prone to increase the local field values. Considering that the experimental structures are more complicated and irregular than the models used in simulation, the realistic enhancement factor will be much higher and will be more beneficial for the detection of tiny amount of materials.

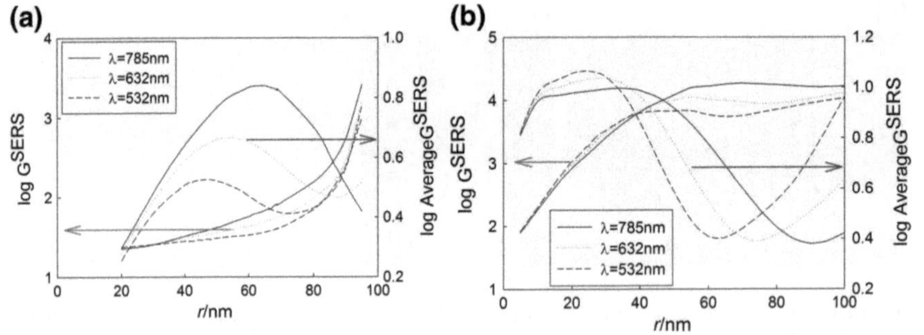

Fig. 14.2 **a** The enhancement factor and average enhancement factor as a function of the NP radius r when the pitch is fixed at 200 nm. **b** The enhancement factor and average enhancement factor as a function of the NPs radius r when the distance d is fixed at 5 nm

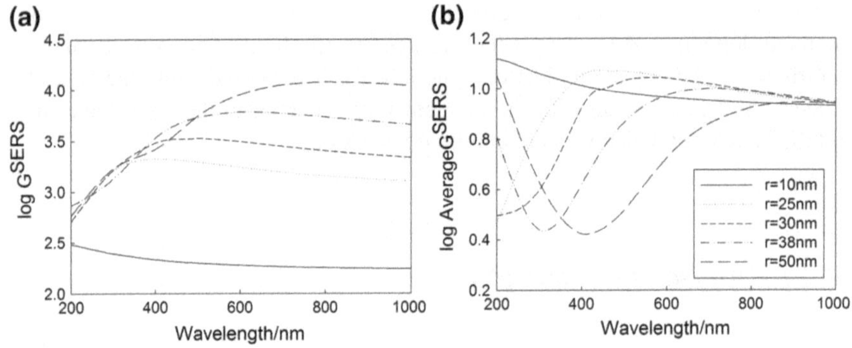

Fig. 14.3 The enhancement factor (**a**) and average enhancement factor (**b**) as a function of the incident wavelength λ when the distance between the NPs was set at 5 nm

With the pitch of the two NPs was fixed at 200 nm, the enhancement factors as a function of the sizes of NPs was shown in Fig. 14.2a. It can be seen that as the radius of the NPs increases, the enhancing factor increases monotonously, while the average enhancing factor has a peak, which shifts to a higher radius as the wavelength increases. When designing, a precise radius should be chosen to ensure both a high enhancing factor and a considerable average enhancing factor, and this is also true when the distance between the NPs is fixed, as shown in Fig. 14.2b.

With the distance between the NPs set at 5 nm, the enhancement factors as a function of the sizes of NPs was shown in Fig. 14.3. It can be seen that there is a peak for each case, indicating resonance between the NPs and the incident wave. And the location and strength of the peak both increases as the radius of the NPs increase, however, at the mean time, the average enhancement factors decrease-still there should be a balance between those parameters.

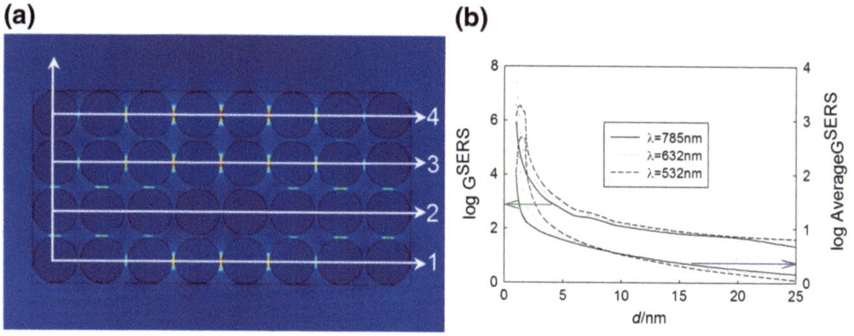

Fig. 14.4 **a** The over-all electric field distribution between arrays of NPs, where d = 1 nm, r = 10 nm, λ = 532 nm. The *white arrows* indicate the centerline of each array. **b** The enhancement factor and average enhancement factor of the NPs array as a function of the distance d

14.2.2 Array of Two NPs

Normally on SERS substrates, the nanostructures will most probably exist in the form of ordered arrays; so consequently, the distance between each nanostructures pairs is also one important issue that must be taken into account. Here ordered, not stochastic arrays were simulated.

Figure 14.4a shows the electric field distribution of the NPs array, with four white arrows indicating the centerline of each array. Apparently the electric distribution across the array is not uniform. Figure 14.4b shows the enhancement factor and average enhancement factor of the NPs array as a function of the distance *d* between the NPs. As the distance *d* increases, both enhancement factors will decrease, as expected.

Though the enhancing factors decreases as the distance increase, it could be clearly seen that as the distance increases from 1 to 10 nm for NPs of 10 nm radius, the uniformity of the enhancing factors gets better. The electric field strengths in the first column in Fig. 14.5 were quite discrete, with a big difference between each peak, meaning that electric field strength along the four centerlines deviate quite much from each other. The strength scattering can also be seen from Fig. 14.4a, where the distance between the NPs pairs were too close (1 nm), the electromagnetic wave along Line 2 is likely 'blocked' by the other NPs, with little enhancement between each NP on this line. The electric field strengths in the last two columns are smoother that the first one obviously.

So when designing a NP patterned SERS substrate, the NP array could not be too dense, which means, the distance between the NPs cannot be too small. However, if the NPs array were too loose, the enhancing effect will deteriorate and also the 'hot spots' would be too sparse, so a balance should be clarified for the NPs arrangement.

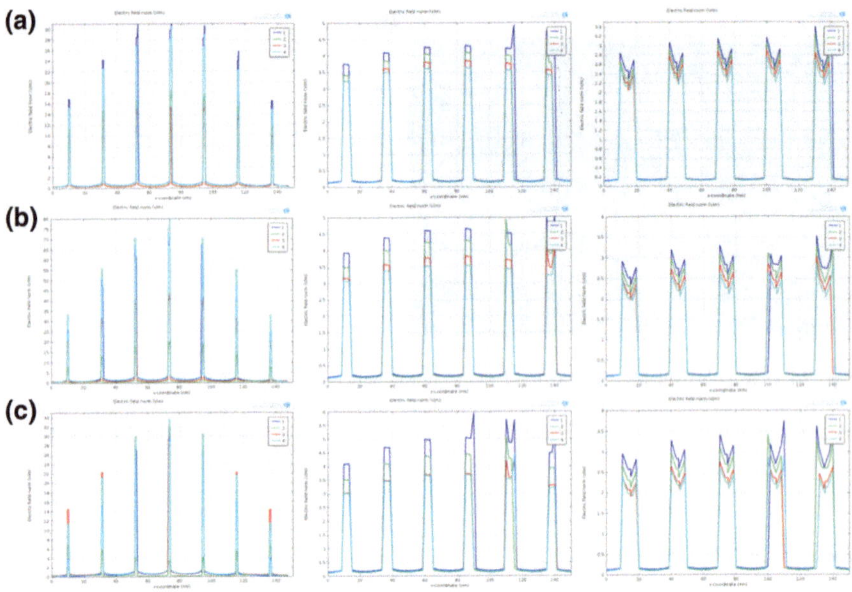

Fig. 14.5 The over-all electric field distribution along the *four centerlines* in Fig. 14.4a, where the radius of NPs is 10 nm. **a** $\lambda = 785$ nm, **b** $\lambda = 632$ nm, **c** $\lambda = 532$ nm. From *left* to *right*, the *three rows* of figure represent cases of d=1 nm, d=5 nm and d=10 nm, respectively

The geometrical parameters of the enhancement factor were demonstrated, and it can be concluded that the simulated results are in large scale in corresponding with published experimental results. And for a better uniform substrate, not only the size, distribution of the nanostructures should be optimized, but also the distance between the nanostructure pairs should be carefully defined.

14.3 Conclusions

To further reveal the physical mechanism of the enhancements of Raman scattering of gold NPs and to optimize the arrangement of NPs, we carried out thorough simulations of local field distributions. Two types of simplified gold NPs models, including smooth a NPs pair and a NP array, were calculated and compared. To both ensure a high maximal enhancement factor and a high average enhancement factor in a fixed domain, a new parameter AverageG was proposed. It can be concluded for a single pair of NPs, bigger NPs will result in a longer resonance wavelength. And as the distance gets longer, both the enhancement factor and the average enhancement factor will decrease dramatically. However, for a whole distribution of NPs on a whole substrate, if they are placed too close, the inner NPs will be largely 'shielded'. So when it comes to a whole substrate, the distance between the NPs should be defined by carefully balancing the enhancement factors and their consistence.

Acknowledgments The work was supported by the National Natural Science Foundation of China (No. 51205245), Science and Technology Commission of Shanghai Municipality (No. 11PJ1403500), and Innovation Program of Shanghai Municipal Education Commission (No. 12YZ022).

References

1. Lim D-K et al (2009) Nanogap-engineerable Raman-active nanodumbbells for single-molecule detection. Nat Mater 9(1):60–67
2. Pieczonka NP, Aroca RF (2008) Single molecule analysis by surfaced-enhanced Raman scattering. Chem Soc Rev 37(5):946–954
3. Tripp RA, Dluhy RA, Zhao Y (2008) Novel nanostructures for SERS biosensing. Nano Today 3(3):31–37
4. Mishchenko MI, Travis LD, Lacis AA (2002) Scattering, absorption, and emission of light by small particles. Cambridge university press, Cambridge
5. Panigrahi S et al (2006) Self-assembly of silver nanoparticles: synthesis, stabilization, optical properties, and application in surface-enhanced Raman scattering. J Phys Chem B 110(27):13436–13444
6. Braun G et al (2007) Surface-enhanced Raman spectroscopy for DNA detection by nanoparticle assembly onto smooth metal films. J Am Chem Soc 129(20):6378–6379
7. Tao A et al (2003) Langmuir-Blodgett silver nanowire monolayers for molecular sensing using surface-enhanced Raman spectroscopy. Nano Lett 3(9):1229–1233
8. Li W-D et al (2011) Three-dimensional cavity nanoantenna coupled plasmonic nanodots for ultrahigh and uniform surface-enhanced Raman scattering over large area. Opt Express 19(5):3925–3936
9. Tian C et al (2011) Nanoparticle attachment on silver corrugated-wire nanoantenna for large increases of surface-enhanced Raman scattering. ACS Nano 5(12):9442–9449
10. Kneipp K et al (1997) Single molecule detection using surface-enhanced Raman scattering (SERS). Phys Rev Lett 78(9):1667–1670

Chapter 15
An Image-Based Coordinate Tracking System Using Afocal Optics for Surgical Navigation

You Seong Chae, Seung Hyun Lee, Min Young Kim and Hyun Ki Lee

Abstract The optical tracker is the equipment for real-time 3D position and pose measurement of marker using camera. The general optical tracker uses the stereo camera and rigid body marker. When using in surgical application, the rigid body marker is attached to the surgical instrument or patient. Therefore, it is necessary to reduce the size of attached marker for the convenience of surgeon. But if the size of the attached marker reduces, the size shown in the camera image also reduces. Then the accuracy of calculated coordinates decreases. To solve this problem, a novel coordinate tracking method is proposed which uses the bokeh phenomenon. The afocal system which uses this method is used with a marker which has a fish-eye lens in front, behind it a 10 mm× 10 mm glass pattern with 30 μm× 30 μm datamatrices printed on. If the camera lens is out of focused, these small data-martices are magnified and shown in the image took by the camera. The tracking method calculates the geometrical relation of the camera and the marker using the information of the patterns. Then the position and pose of the marker are calculated using the result of geometrical relations. Because the patterns are magnified, the accuracy of measurement is high. Also for the size of glass pattern is small, the total size of the marker also can be reduced. The result of experiments shows that the proposed algorithm is usable in image based surgical navigation.

Keywords Afocal optics • Infinity focus • Stereo camera • Optical tracker • Pose tracking • Pattern matching

---

Y. S. Chae · S. H. Lee · M. Y. Kim (✉)
School of Electronics Engineering, Kyungpook National University, Daegu, South Korea
e-mail: mykim@ee.knu.ac.kr

Y. S. Chae
e-mail: ysch@ee.knu.ac.kr

S. H. Lee
e-mail: shlee77@ee.knu.ac.kr

H. K. Lee
Medical Robot Unit, Kohyoung Technology, Seoul, South Korea
e-mail: hklee@kohyoung.com

R. Tutsch et al. (eds.), *Progress in Optomechatronic Technologies*,
Lecture Notes in Electrical Engineering 306, DOI: 10.1007/978-3-319-05711-8_15,

15.1 Introduction

Recently, image based surgical robot system has been rapidly developed resultant of the development of computer and optical technology [1]. Due to the application of image guided system on robot surgery, sensor based accurate robot control, and tracking of the positions of surgical tools and patient, navigation of robot surgery is now capable. General image based optical tracking system uses a rigid body marker with retro-reflective spheres attached to it. The system performs tracking by calculating the position and pose of the marker through geometric analysis of the marker spheres using stereo camera [2]. In comparison with the old days, the size of the rigid body marker is small. However, it quite disturbs the user when it is attached to a small tool, especially when applying it in elaborate work as surgery. If rigid body marker attached to the patient or surgical instruments overlaps each other, the identification of each rigid body marker is difficult. Also as the distance between the marker and camera becomes distant, the accuracy of measurement of position and pose decreases. Because most of the tracking systems use similar methods, the tracking systems all have these kinds of problems. Because the basic technique is similar to each other, it is hard to improve the weaknesses of the optical tracker. But surgical robot systems require a more accurate tracking system to improve the performance and safety. Therefore a new method for tracking system that can overcome these weaknesses is required, and this is introduced in this paper.

The idea came from the bokeh phenomenon that is used in bokode [3]. Bokode uses the bokeh phenomenon by using two separate optical lenses. A lenslet is in front of pattern that contains datamatrices and the camera observes this with the camera lens out of focused. The bokode pattern appear as tiny dots to the human eye and camera in sharp focus, but appears much bigger to the out of focused camera. Therefore, the pattern can be identified accurately even with a distance of few meters. The advantage of the bokode is that it can contain much more data in smaller area compared to barcodes or other tags. There is an application using bokode to measure position and angle of the bokode too. However, the method is not accurate and the position and angle are dependent to each other. Which means it measures the relative position and angle. Therefore it can't be used as a tracking method.

In this paper, afocal tracking system that uses the bokeh phenomenon [4] and the pattern is introduced. The system has two cameras as a sensor, an afocal marker to be attached on tracking target. The afocal tracking system measures the absolute position and pose of the marker, which means it is possible to use it as a tracking system. The system can use smaller markers than traditional optical trackers and can maintain the accuracy of measurement independent of distance. Also for this method is a new one, it has strong possibility to be improved.

15.2 Method

15.2.1 Basic Idea

Before details of the measurement method, understanding of the bokode is desirable. It is based on the bokeh effect of ordinary camera lenses. The concept of bokode is to use more data with a smaller code pattern. Also bokode can read the data of the codes from farther distances. But the most important point we focused on was that the patterns in the bokeh changes by the position and angle of the bokode. But in the study, the pattern fades as the angle between the camera and the bokode increases by lens aberration. In the bokode study, there is an application for measuring the angle and the position of the bokode. But the position and angle measured are related to each other, so if the position and the angle changes at the same time, it can't know the real position and rotation of the bokode. For example, if the position of the bokode changes without rotation, the measured angle changes too because the measured angle is just the angle between the camera lens optical axis and the bokode lenslet optical axis. Also the depth is measured by the diameter of the diameter of the bokeh, but the change rate of the diameter of the bokeh related to the depth is small, so the measuring resolution is low. Though, the study of the bokode has big significance in tracking because it is a new method of measuring positions and angles. In this paper, a tracking method using the concept of the bokeh to measure the accurate position and the accurate pose of the marker is introduced. We call it an afocal tracking system.

The difference of the measuring method is that afocal tracking system measures the absolute position and pose of the marker using two cameras. Also the viewable angle is improved by using fisheye lens in front of the marker pattern. It will be mentioned as afocal marker.

Before configuring the system, three methods of tracking was considered. The three methods are shown in Fig. 15.1. One is one camera with stereo afocal marker, another is stereo camera with stereo afocal marker, and the other is stereo camera with one afocal marker. System (A) has an advantage in cost because it requires only one camera, but has a disadvantage that the marker size is doubled for it uses two afocal markers. System (B) has an advantage in accuracy. The afocal tracking system has better accuracy of measuring the vertical and horizontal measurement than the depth measurement. System (B) calculates the depth using the vertical and horizontal measurement therefore performs higher accuracy in measuring the depth. But it has a disadvantage of cost and size of marker for it uses two cameras and two afocal marker. System (C) has an advantage of the marker size for it uses one afocal marker. The cost is higher than system (A) but lower than system (B). Considering surgical application, which is important to use a small target marker for convenience of surgery, System (C) was applied and developed.

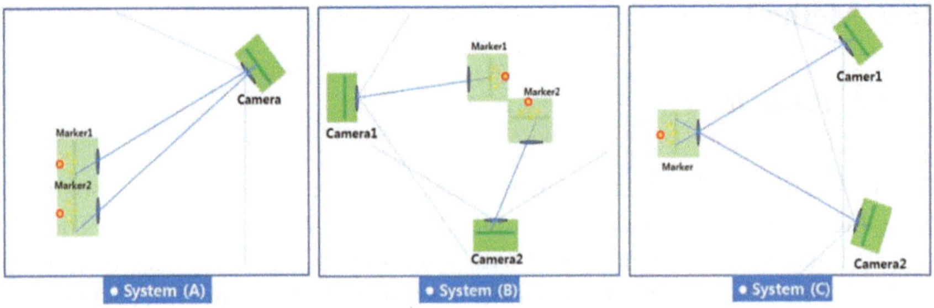

Fig. 15.1 Configurations of tracking systems for afocal tracking system [(A) One camera with stereo afocal marker, (B) Stereo camera with stereo afocal marker, (C) Stereo camera with one afocal marker]

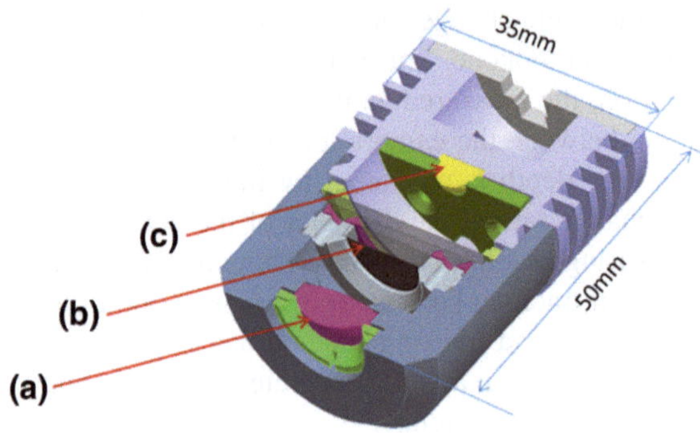

Fig. 15.2 Concept of afocal marker for afocal tracking system (*a* Fisheye lens. *b* Glass pattern. *c* Infra-red LED)

15.2.2 Hardware

The hardware configuration of afocal tracking system is very important to perform accurate tracking in proper area. The afocal marker has a fisheye lens in front, a glass pattern with datamarices printed on, and an infrared LED behind the glass pattern. The design of the afocal marker is shown in Fig. 15.2.

The reason of using infrared light is so that we can improve the accuracy and speed of image processing. There is a infra-red filter in front of the camera lens so that the images obtained from cameras only includes the bokeh and the pattern from the afocal marker. To do tracking of proper tracking area, the aperture of the camera lenses, the focal length of the camera lenses, the focal length of the fisheye lens, and the geometrical position of the two cameras has to be carefully modulated. Also the printing size of the pattern on the glass pattern and the pixel size of the camera CCD has to be considered for it decides the resolution of pattern in the obtained images. The resolution has to be high enough to decode the datamatrices. In this study, for high resolution we used a 1 mm by 1 mm sized pattern glass that has 91,809 datamatrices printed on it. A photomicrograph of data matrix printed on glass pattern is shown in Fig. 15.3.

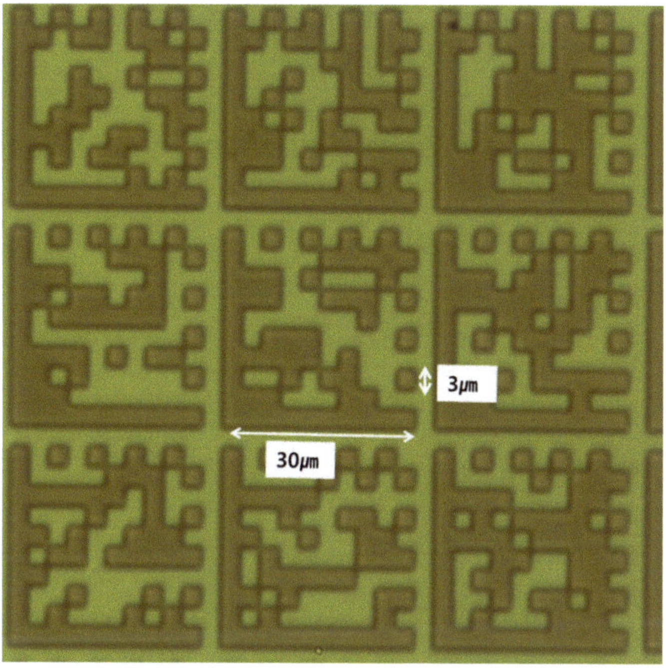

Fig. 15.3 Photomicrograph of data matrices printed on glass pattern

The developed in this application is configured for FOV (Field Of View) of 200 mm by 200 mm in 300 mm distance from the stereo camera. The system considering the FOV is shown in Fig. 15.4.

15.2.3 Measuring of Position and Pose of Afocal Marker

For position measuring, two methods are developed. The first one is based on stereo camera system and the second one is using the geometrical relations of patterns and camera CCD.

To use stereo camera system in afocal tracking system, calibration is somewhat difficult. Because the camera lenses of afocal tracking system are set to out of focus, ordinary camera calibration using chessboards can't be used. Therefore a calibration board is designed with corner points formed by pinholes and LED light illuminating behind it. The LED light goes through the pinholes, blurs in the image as a circle. Through circle fitting, the center points of these blurred circles are found and Zhang's calibration [5]. But this calibration board is not yet manufactured. Instead for temporary method linear motor stages are used to change the afocal marker's position to position it on all calibration center points, and the center points are extracted by finding the center of the bokeh. After calibration, measuring the position of the afocal marker is done by stereo camera system. Then

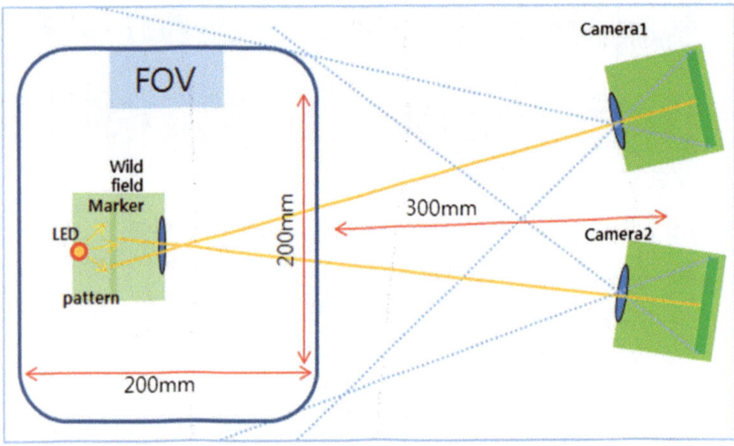

Fig. 15.4 Tracking system considering the FOV

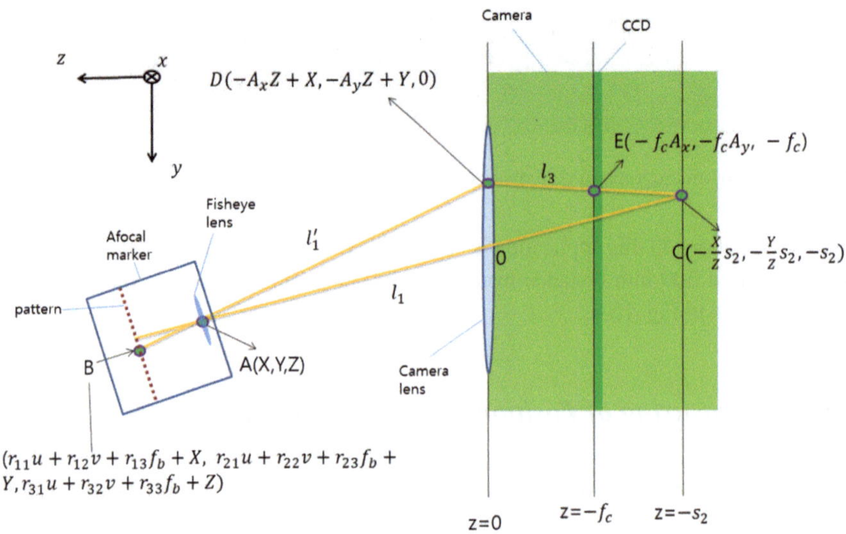

Fig. 15.5 Geometrical relation of afocal marker and camera

measuring the pose of the afocal marker is done by the geometrical relation of one afocal marker and the camera which is shown in Fig. 15.5. The geometrical relation can be expressed as equation.

$$S\begin{bmatrix} u' \\ v' \\ 1 \end{bmatrix} = [A][R][C]\begin{bmatrix} u \\ v \\ 1 \end{bmatrix} = \begin{bmatrix} -\frac{f_c}{pw.} & 0 & u_c' \\ 0 & -\frac{f_c}{ph.} & v_c' \\ 0 & 0 & 1 \end{bmatrix}\begin{bmatrix} r_{11} & r_{12} & r_{13} \\ r_{21} & r_{22} & r_{23} \\ r_{31} & r_{32} & r_{33} \end{bmatrix}\begin{bmatrix} 1 & 0 & -u_c \\ 0 & 1 & -v_c \\ 0 & 0 & f_b \end{bmatrix}\begin{bmatrix} u \\ v \\ 1 \end{bmatrix}$$
$$(s = r_{31}u + r_{32}v + r_{33}f_b) \tag{15.1}$$

Equation (15.1) is similar to that of Zhang's calibration [5]. By obtaining a few images of the afocal marker with random position and poses, and then using similar calculation of Zhang's calibration we can know the matrix H, A and C. with Eq. (15.2) the pose of the afocal marker is obtained.

$$[H] = [A][R][C] \tag{15.2}$$

$$[R] = [A]^{-1}[H][C]^{-1} \tag{15.3}$$

Another method is using the geometrical relations of the afocal tracking system. This method measures the position and the pose simultaneously. The geometrical relations of afocal tracking system are shown in Fig. 15.6. For better understanding, the method to obtain the position and pose is described in two-dimensional coordinate. The small lens (fisheye lens of afocal marker) and the dotted line (glass pattern of afocal marker) are rotated from 0° to the positive direction with θ. L1 and L2 is the center of the camera lenses of camera1 and camera2. P1 is the position of the datamatrix shown in the center of the bokeh of camera1 and C1 is the position of the bokeh center CCD pixel of camera1. P2 is the position of the datamatrix shown in the center of the bokeh of camera2 and C2 is the position of the bokeh center CCD pixel of camera2.

The center of the fisheye lens is (X, Y), Point P1 is $(f_f \cdot \cos\theta - u'_2 \cdot \sin\theta + X, \sin\theta \cdot f_f + u'_2 \cdot \cos\theta + Y)$, point P2 is $(f_f \cdot \cos\theta - u'_1 \cdot \sin\theta + X, \sin\theta \cdot f_f + u'_1 \cdot \cos\theta + Y)$, L1 is (0, 0), L2 is (0, L), C1 is $(-f_c, -u_1)$ and C2 is $(-f_c, -u_2)$. f_c is the focal length of the camera lens, f_f is the focal length of the fisheye lens, u_1 is the bokeh center pixel of camera1, u_2 is the bokeh center pixel of camera1, u'_1 is the distance between the glass pattern center and the datamatrix at point P1, u'_2 is the distance between the glass pattern center and the datamatrix at point P2 and L is the distance between the two camera lenses. Point P1, the center of the fisheye lens, L1 and C1 are on line1. Point P2, the center of the fisheye lens, L2 and C2 are on line2. Therefore equations of straight line made of the four points on line1 equals each other, and the equations of straight line made by the four points on line2 equals each other. Solving the system of equations, the angle θ, X and Y are obtained.

$$\tan\theta = \frac{-f_f(u_2 - u_1) - f_c(u'_2 - u'_1)}{u'_1 u_1 - u'_2 u_2} \tag{15.4}$$

$$X = \frac{\left[(u_1 + u_2)f_f - (u'_1 - u'_2)f_c\right]\cos\theta - (u'_1 u_1 - u'_2 u_2)\sin\theta - Lf_c}{u_1 - u_2} \tag{15.5}$$

$$Y = \frac{\left[(u'_1 u_2 - u'_2 u_1)f_c - 2u_1 u_2 f_f\right]\cos\theta + \left[(u'_1 + u'_2)u_1 u_2 - (u_1 + u_2)f_f f_c\right]\sin\theta + Lf_c u_1}{(u_1 - u_2)f_c} \tag{15.6}$$

Doing the same process in three-dimensional coordinates, the position and the pose of the afocal marker are obtained.

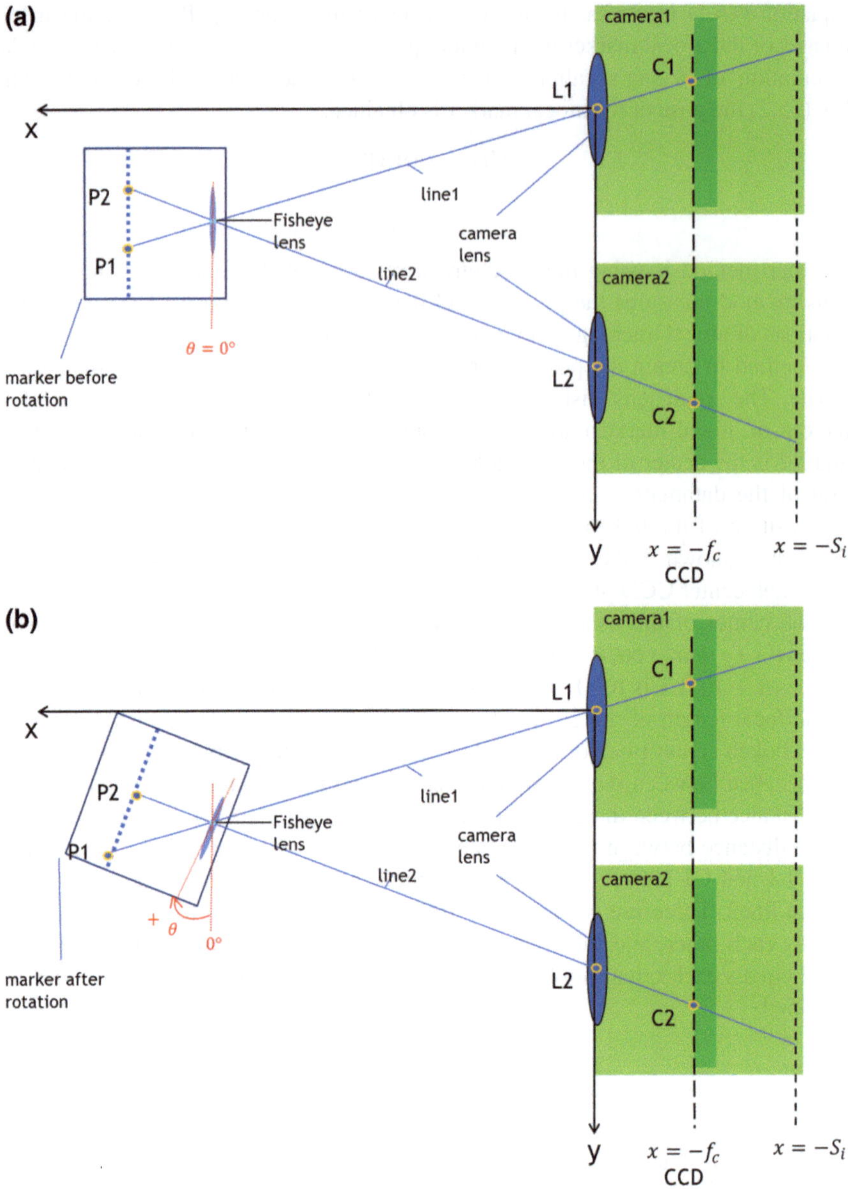

Fig. 15.6 Geometrical relations in afocal tracking system (**a** when there is no rotation of the afocal marker and **b** when there is rotation of the afocal marker)

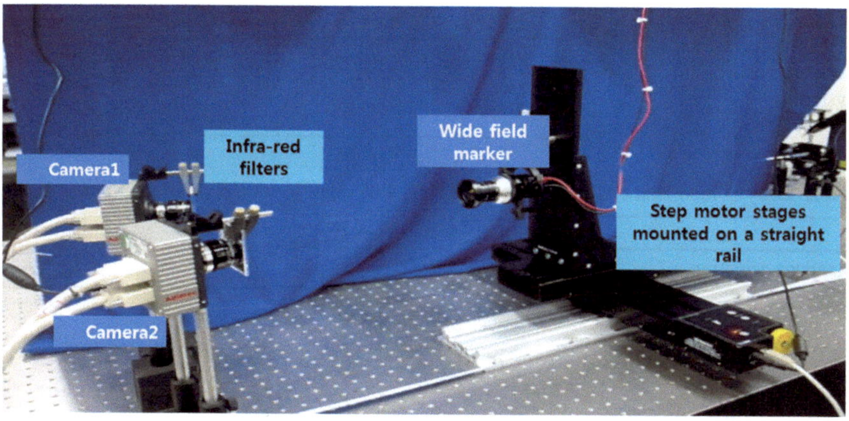

Fig. 15.7 The experiment for position measurement

15.3 Experiment

15.3.1 Position Measurement

Experiments are done for the first method which is proposed in Sect. 15.2. The experimental setup for the second method is in progress and the experiment and analysis is left for further work.

To do this experiment, a step motor stage was used to change the position of the afocal marker. First, two step-motor stages were mounted for the vertical and horizontal direction on a straight rail which was position for the depth direction as shown in Fig. 15.7. Then the vertical and horizontal direction position was measured by afocal tracking system with various depth by changing the depth position with the rail.

After that the two step motor stage was mounted for the vertical and depth direction on a straight rail which was position for the horizontal direction. Then the depth position was measured by afocal tracking system with various horizontal and vertical direction. The measurement of each direction was done 150 times. Error between real moving distance of marker using motor stage and measured moving distance of marker using afocal system is calculated. And standard deviation of calculated error is shown in Table 15.1. The 'y' for the vertical direction, 'x' for the horizontal direction and 'z' for the depth direction. The comparison of performance with commercialized optical tracker named 'polaris vicra' of NDI corp. (http://www.ndigital.com) is shown in Table 15.2.

In case of afocal system, average position error is 200 μm. And standard deviation of error is 219 μm. But standard deviation of position error is 452 μm by using Polaris vicar. So it proves that afocal system measures more accurately better than Polaris vicar.

Table 15.1 Standard deviation of position measurements of afocal system

Coordinate	Average error (μm)	Standard deviation (μm)
x	197	210
y	213	182
z	242	265

Table 15.2 Comparison of position measurements with commercialized optical tracker

	Afocal system	Polaris vicra
Standard deviation	219 μm	452 μm

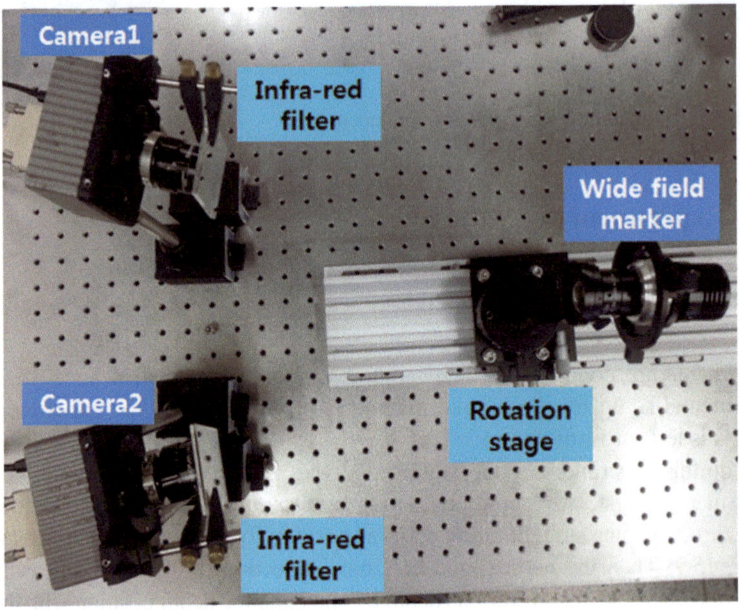

Fig. 15.8 The experiment for pose measurement

15.3.2 Pose Measurement

To do this experiment, a rotation stage was used as shown in Fig. 15.8. As it was difficult to make a roll, pitch and yaw controlling stage, the measurement experiment was done in one directional rotation axis. The rotation stage was rotated from 11° to −15° with every 1° increment in x-axis. As the manual rotation stage was operated by hand, the rotational accuracy of the stage was assumed as almost 0.03°. The results are shown in Table 15.3 and Fig. 15.9.

In Fig. 15.9, because only x axis is rotated by the manual rotation stage, values of y and z axis are almost steady. The rotation angle of x axis is changed by 1° increment consistently at every sensing time, and the sensing tests are performed in every change.

Table 15.3 Results of pose measurements

	Average error	Standard deviation
X axis measurement	0.022°	0.151°
Y axis measurement	0.198°	0.229°
Z axis measurement	0.021°	0.027°

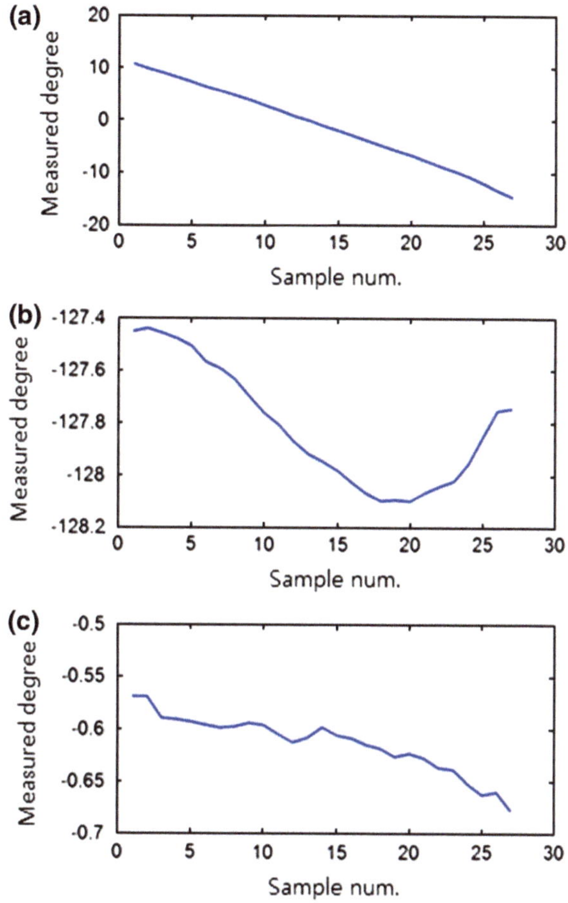

Fig. 15.9 Result of pose measurement with x-axis rotation. **a** X-axis. **b** Y-axis. **c** Z-axis

Average error and error standard deviation of the rotation measurements show 0.02° and 0.15° in x-axis. However, the results in y-axis show somewhat large error. It results from the misalignment between coordinates of the sensor and the rotation stage.

The obtained result of position and pose measurement shows that the proposed system is tracking in a reasonable accuracy. In the case of pose measurement, it uses only one camera for calculation. The accuracy of pose measurement will surely be improved using two cameras and this is in progress. After the whole system is implemented, experiments will be done to compare the accuracy with other commercial tracking systems [6] and inspect that the proposed system is usable for surgical application.

15.4 Conclusion

Expectation and demand of surgical robot application are steadily increasing. Therefore accurate image based robot controlling for safe and accurate robot surgery is becoming more and more important. Also, the size of markers which are attached to patients or surgical instruments needs to be smaller for convenience of surgery. Therefore in this paper, a novel method of tracking that is possible to miniaturize the marker and has accurate measurement of position and pose is proposed. Afocal tracking system also maintains the measurement accuracy independent of the distance of the target. Although, manufacturing a smaller sized afocal marker is left for further work. The fisheye lens used in the proposed system is a commercial lens and it takes almost half of the size of the afocal marker. The development of a much smaller fisheye lens that can be mounted in the afocal marker is in progress. The accuracy of pose measurement will surely be improved after using two cameras for measurement. The importance of this work is that it is a novel method of tracking, and therefore has strong possibility of improvement.

Acknowledgments This work is supported by the Technology Innovation Program (10040097) funded by the Ministry of Knowledge Economy (MKE, Korea).

References

1. Rasool K, Clement CY, Mohammad ST, Michael RB, Jeremy AJ, Jacqueline NW, Eric PW, Ramin S (2000) Comparative tracking error analysis of five different optical tracking systems. Comput Aided Surg 5:98–107
2. Pintaric T, Kaufmann H (2007) Affordable infrared-optical pose-tracking for virtual and augmented reality. In: IEEE VR workshop on trends and issues in tracking for virtual environments. IEEE Press, New York, pp 44–51
3. Mohan A, Woo G, Hiura S, Smithwick Q, Raskar R (2009) Bokode: imperceptible visual tags for camera based interaction from a distance. ACM Trans Graph 28:98
4. Wu J, Zheng C, Hu X, Wang Y, Zhang L (2010) Realistic rendering of bokeh effect based on optical aberrations. Visual Comput 26:555–563
5. Zhang Z (2000) A flexible new technique for camera calibration. IEEE Trans Pattern Anal Mach Intell 22:1330–1334
6. Chae YS, Lee SH, Kim MY (2012) Multiple optical system for matching coordinate system of neurosurgical robot. In: The 9th international conference on ubiquitous robots and ambient intelligence. IEEE Press, New York, pp 75–79

Chapter 16
Feasibility Study of a Functional Near Infrared Spectroscopy as a Brain Optical Imaging Modality for Rehabilitation Medicine

Seung Hyun Lee, Sang Hyeon Jin, Jinung An, Gwanghee Jang, Hyunju Lee and Jeon-Il Moon

Abstract A functional near-infrared spectroscopy (fNIRS)—which is a non-invasive modality to measure hemodynamics of cortices—is today the frequently reported optical brain imaging method comparing with fMRI from the standpoint of clinical feasibility. The aim of this study was to explore the experience of fNIRS in several motor executions which cannot be implemented in an fMRI experimental condition, to describe their cortical activations, and consequently to examine the feasibility of fNIRS as an acceptable brain imaging technology for rehabilitation medicine. Five healthy men performed the individually given tasks. Five tasks were offered in this study: active hand grasping (hand flexion and extension), active arm raising (shoulder flexion and extension), active eating (ADL task of upper extremity), active knee bending (leg flexion and extension in sliding bench), and active walking (ADL task of lower extremity in treadmill). The fNIRS cortical map of each tasks coincided with the cortical areas where should be activated at each motor functions shown in many related literatures approving the same neurophysiological fact by fMRI or other brain imaging modalities. The results from this study may contribute to better understanding how motor executions can be expressed into cortical activation patterns via fNIRS measurement. The ability of fNIRS to image cortical activations at satisfactory spatiotemporal resolutions makes fNIRS a potentially powerful non-invasive brain imaging modality for diagnosis and evaluation of the motor performance for patients in rehabilitation medicine.

Keywords fNIRS • Optical brain imaging • Cortical activation • Rehabilitation

S. H. Lee · S. H. Jin · J. An (✉) · G. Jang · H. Lee · J.-I. Moon
Robotics Research Division, DGIST, Daegu, Republic of Korea
e-mail: robot@dgist.ac.kr

S. H. Lee
e-mail: aksska82@dgist.ac.kr

R. Tutsch et al. (eds.), *Progress in Optomechatronic Technologies*,
Lecture Notes in Electrical Engineering 306, DOI: 10.1007/978-3-319-05711-8_16,

16.1 Introduction

In recent years, a new technique for brain mapping, functional near-infrared spectroscopy (fNIRS) [1], has been gradually spread out. Promising methods in neuro-muscular rehabilitation have been introduced from the sufficient evidence of neural plasticity, which refers to the ability of the brain to change its structure and function by external stimuli [2].

Several studies have applied functional magnetic resonance imaging (fMRI) and positron emission tomography (PET) to study neuronal activation in the rehabilitation tasks [3]. They are powerful tools of brain mapping for somatosensory and cognitive tasks as well as motor tasks of hand [4]. However, fMRI and PET are also present several challenges such as high sensitivity to participant motion, a loud, restrictive environment, and relatively high cost. Some of these challenges are overcome with fNIRS.

The purpose of this study is to investigate the feasibility of utilizing fNIRS in the field of rehabilitation medicine through recording brain activation during some motor function tasks. fNIRS, similar to fMRI, is a non-invasive brain imaging technique that can measure blood oxygenation changes during neural activities [5, 6]. fNIRS, presents several advantages relative to fMRI, such as a small, portable, relatively safe, and low cost method of monitoring of brain activity [7]. fNIRS can be quite suitable to study in the motor function task because it can measure brain activations on allowing subject's motion [8, 9]. Thus, we use the fNIRS to analyze the cortical activity during several motor function tasks, which are widely adapted in clinical rehabilitation treatments: hand grasping, arm raising, eating, knee bending, and walking.

16.2 Materials and Methods

16.2.1 Task and Procedure

Total five healthy right-handed volunteers between 29 and 32 ages (6 males) without known neurological, physical abnormalities participated in this study. All subjects understood the purpose of the study and gave the written informed consent prior to participation.

Five tasks were performed in this paper. (1) Active hand grasping (hand flexion and extension): participant was instructed to flex and extend their right hand with a frequency of 0.5 Hz paced by an auditory cue presented via digital metronome. (2) Active arm raising (shoulder flexion and extension): participant laid supine on a bed and he instructed to flex and extend his arm with a frequency of 0.33 Hz paced by an auditory cue presented via digital metronome. (3) Active eating (an ADL task of upper extremity): subject was instructed to transfer an ice-cream with spoon from plate to their mouth, taste, and swallow it with a frequency of 0.5 Hz paced by an auditory cue presented via digital metronome. (4) Active

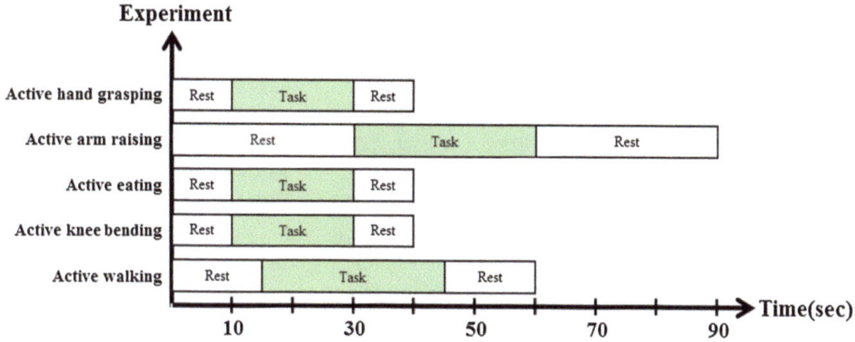

Fig. 16.1 The experimental protocols of five tasks

knee bending (leg flexion and extension in sliding bench): volunteer was laid on a bed in supine position, and he instructed to flexion and extension his leg with a frequency of 0.5 Hz paced by an auditory cue presented via digital metronome; (5) Active walking (an ADL task of lower extremity): volunteer walked at 3 km/s on a treadmill periodically.

Each protocol of five tasks was shown in Fig. 16.1. All protocols were implemented as block schemed experiments. Every task was repeated three or five times. Tasks and rests were cued by beep sound at every start and stop. In rest phase, subject was instructed to relax without performing any movements and imagination.

16.2.2 Optical Brain Imaging Procedure and Data Analysis

The NIRS system (FOIRE-3000, Shimadzu Co., Kyoto, Japan) with wavelength of 780, 805, and 830 nm was used to calculate the cortical activities with a sampling rate of 14 Hz [10]. The system can detect changes in oxygenated hemoglobin (HbO), deoxygenated hemoglobin (HbR), and total hemoglobin (HbT). We made 31 channel system with 20 optodes, consisting of 10 light source and 10 detectors, was used and the inter-optode distance of 30 mm. The optodes were placed on the parietal lobe where covered the primary cortex, somatosensory cortex, and premotor cortex based on 10–20 international electrode placement system. Locations of optodes were measured using a 3D position measuring system (FASTRAC, Polhemus, USA).

The measured data were analyzed using NIRS-statistical parametric mapping (NIRS-SPM). SPM t-statistic maps were computed with the level of significance set at a p-value of <0.01 [11]. On fNIRS analysis, we used the HbO levels as markers of cortical activity because HbO is the most sensitive indicator of changes in regional cerebral blood flow. The obtained images depicted changes in hemoglobin oxygenation of the cortex. In order to investigate the changes in HbO, three regions of interest (ROI) were selected on the basis of the Brodmann area (BA): the primary sensory-motor cortex (BA 1, 2, 3, and 4), the premotor cortex (BA 6),

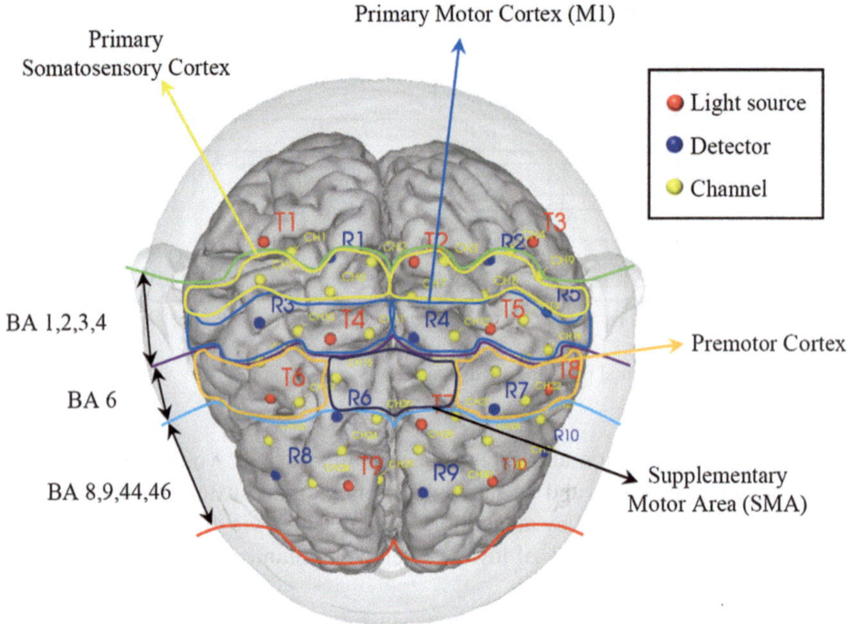

Fig. 16.2 Location of the NIRS optodes and their corresponding cortical regions

and the prefrontal cortex (BA 8, 9, 44, 45, and 46) [12]. The ROIs overlaid on the brain activation image. Figure 16.2 showed the optodes location and the ROIs.

16.3 Results

Figures 16.3 and 16.4 showed the experimental data and analysis results during motor executions of upper extremity, i.e. active hand grasping, active arm raising, and active eating. Figure 16.3 illustrated the extracted examples of time series data of concentration changes of HbO. From this figure, HbO levels increased significantly in task periods comparing with those in rest periods. These results explained that the HbO levels could be an effective indicator to investigate the neural activities for motor execution of upper extremity.

Figure 16.4 represented the results of NIRS topographic mapping for HbO levels which indicated activation of the primary motor cortex (M1), primary sensory-motor cortex (SM1), premotor cortex (PMC), and prefrontal cortex (PFC) during the motor execution of upper extremity, i.e. active hand grasping, active arm raising, passive hand grasping, and active eating. (1) Active hand grasping revealed activation in the M1, contralateral SM1 hand area during flexion and extension of the right hand. (2) During active arm raising, the medial part of M1 and SM1 (trunk and shoulder area) were activated. (3) When eating an ice cream, a

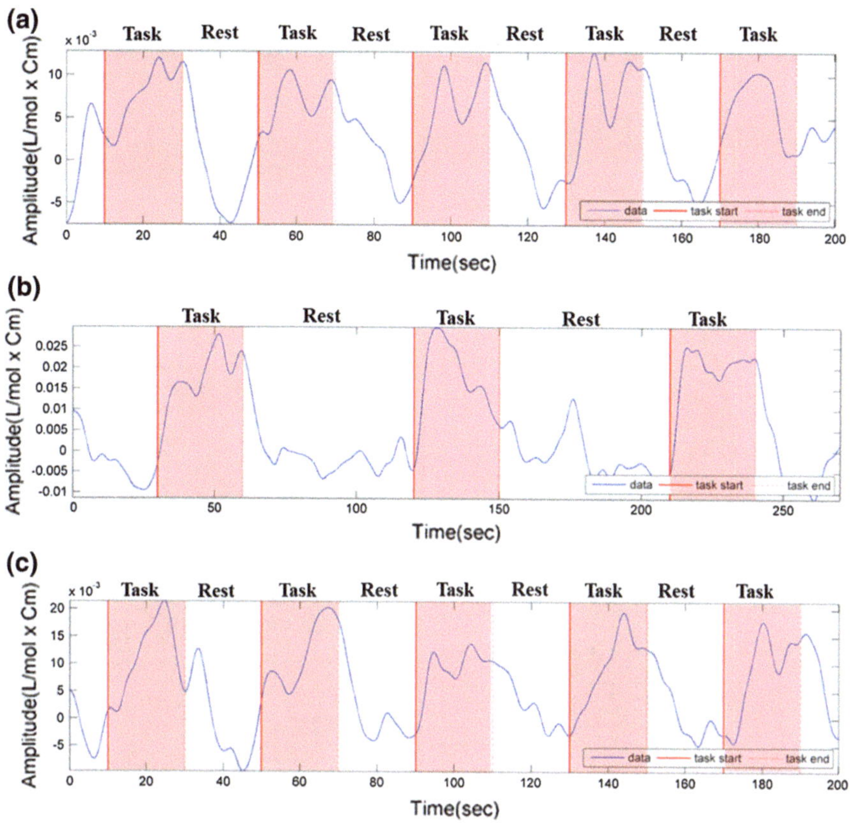

Fig. 16.3 Extracted data of concentration changes of HbO during the motor tasks of upper extremity. **a** Active hand grasping, **b** active arm raising, and **c** active eating

significant increase in HbO levels was mainly observed in the lateral portion of the primary motor cortex and primary sensorimotor regions. The results showed that each motor execution of upper extremity activated the expected sensorimotor network of brain areas corresponding to each motor task.

Figures 16.5 and 16.6 showed the experimental data and analysis results during motor executions of lower extremity, i.e. active knee bending and active walking. Figure 16.5 illustrated the extracted examples of time series data of concentration changes of HbO. Figure 16.5, similar to Fig. 16.3 as the upper extremity case, expressed that HbO levels increased significantly in task periods comparing with those in rest periods. These result described that the HbO levels should be a reliable marker to discover the neural activities for motor execution of lower extremity. Figure 16.6 represented the results of NIRS topographic mapping for HbO levels during the motor execution of lower extremity, i.e. active knee bending and active walking. (4) For the active knee bending, medial primary sensorimotor cortices (lower limb and trunk area) activated. (5) Cerebral activities during active walking

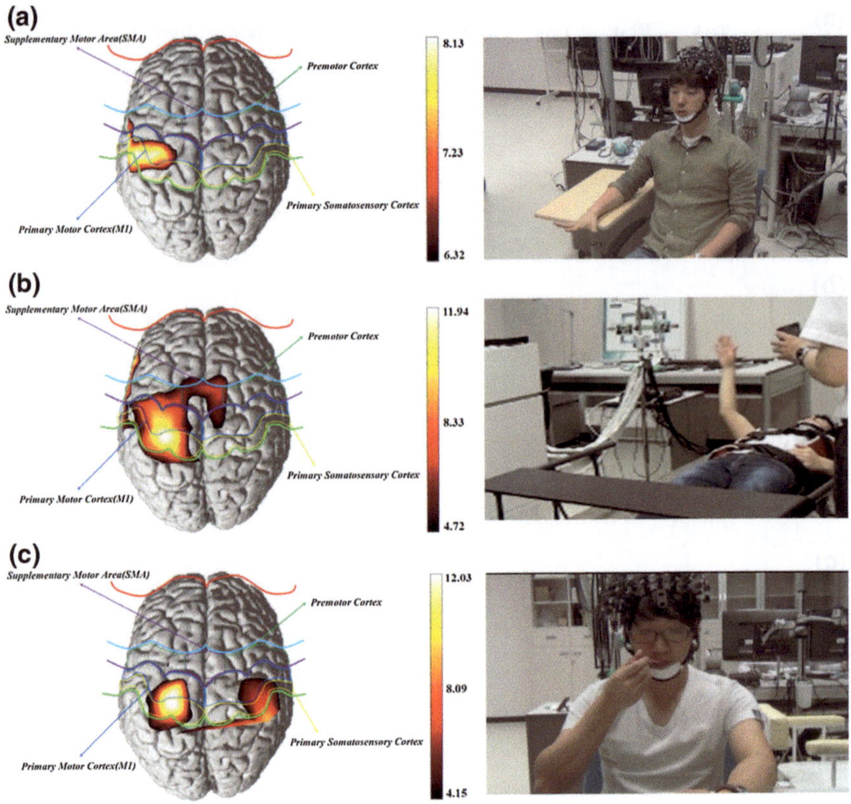

Fig. 16.4 fNIRS cortical activation patterns of motor executions during the depicted upper limb movement tasks. The level of significance was set at a p-value < 0.01. **a** Active hand grasping, **b** active arm raising, and **c** active eating

on the treadmill were located the medial M1 and SM1 (lower limb and trunk area). And also, the supplementary motor area and prefrontal cortex were activated. Walking was predominantly associated with primary motor, somatosensory areas and slightly with the SMA and PFC observed. These two tasks were considered as the rehabilitation of lower extremity. The results claimed that each motor execution of lower extremity activated the expected sensorimotor network of brain areas corresponding to each motor task. Through the experiments, we found the fNIRs is an effective tool for monitoring spectral signatures of brain activity.

16.4 Discussion

This paper was aimed at examining whether fNIRS can be an appropriate brain imaging modality for checking the progress of rehabilitation treatments or not. The five motor tasks commonly used in rehabilitation medicine were chosen to

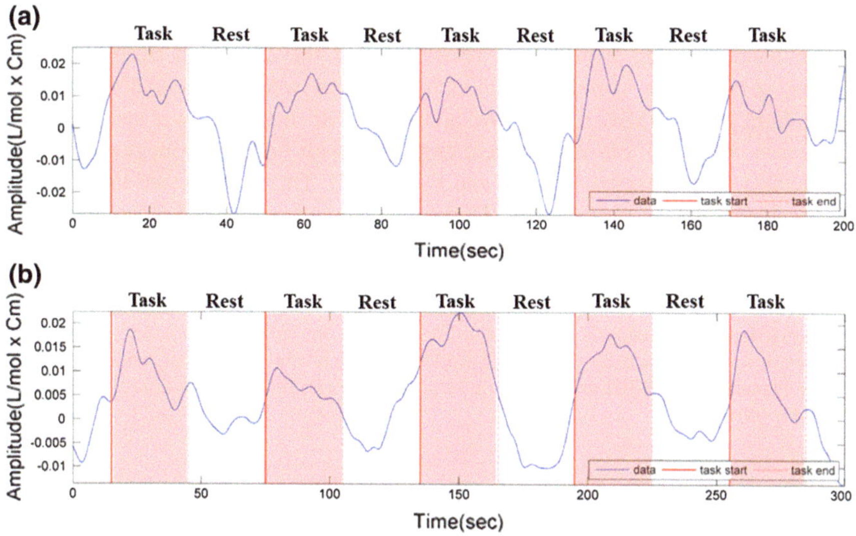

Fig. 16.5 Extracted data of concentration changes of HbO during the motor tasks of lower extremity. **a** Active knee bending and **b** active walking

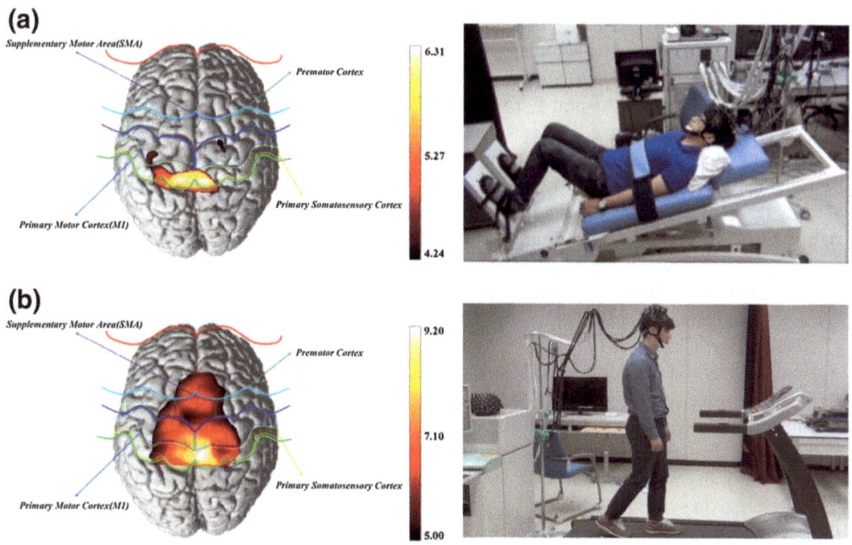

Fig. 16.6 fNIRS cortical activation patterns of motor executions during the depicted lower limb movement tasks. The level of significance was set at a p-value < 0.01. **a** Active knee bending and **b** active walking

validate the feasibility of fNIRS. From upper limb tasks (active hand grasping and arm raising, and active eating) and lower limb tasks (active knee bending and active walking), fNIRS faithfully pointed the brain activity coincided with neurophysiological evidences which were commonly accepted. As obvious evidence, active hand grasping experiments remarked that the cortical activation patterns on the sensorimotor cortex from obtaining fNIRS were considerably similar to those from getting fMRI. Up to now, although NIRS has had lower spatiotemporal resolution than MRI, our results were consistent with findings from several studies comparing signal changes in fMRI and fNIRS [13]. There were some limitations to be considered for the further work. The participants were rather small not to confirm the statistical evaluation of the fNIRS imaging results. But we saw the possibility of the utilizing NIRS into the field of rehabilitation medicine.

Acknowledgments Seung Hyun Lee and Sang Hyeon Jin equally contributed to this work. This work was supported by the DGIST R&D Program of the Ministry of Science, ICT and Future Planning (13-RS-01) and the Korea Ministry of Trade, Industry, and Energy under Grant 2013-45, 10045164.

References

1. Aqil M, Hong K-S, Jeong MY, Ge SS (2012) Cortical brain imaging by adaptive filtering of NIRS signals. Neurosci Lett 514:35–41
2. Kaplan MS (1988) Plasticity after brain lesions: contemporary concepts. Arch Phys Med Rehabil 69:984–991
3. Kwong KK, Belliveau JW, Chesler DA, Goldberg IE, Weisskoff RM, Poncelet BP et al (1992) Dynamic magnetic resonance imaging of human brain activity during primary sensory stimulation. In: PNAS USA, vol 89, pp 5675–5679
4. Takahashi K, Ogata S, Atsumi Y, Yamamoto R, Shiotsuka S, Maki A, Yamashita Y, Yamamoto T, Koizumi H, Hirasawa H, Igawa M (2000) Activation of the visual cortex imaged by 24-channel near-infrared spectroscopy. J Biomed Opt 5:93–96
5. Arenth PM, Ricker JH, Schultheis MT (2007) Applications of functional near-infrared spectroscopy (fNIRS) to neurorehabilitation of cognitive disabilities. Clin Neuropsychologist 21:38–57
6. Calautti C, Baron JC (2003) Functional neuroimaging studies of motor recovery after stroke in adults: a review. Stroke 34(6):1553–1566
7. Strangman G, Culver JP, Thompson JH, Boas DA (2002) A quantitative comparison of simultaneous BOLD fMRI and NIRS recordings during functional brain activation. NeuroImage 17:719–731
8. Hara Y, Obayashi S, Tsujiuchi K, Muraoka Y (2013) The effects of electromyography-controlled functional electrical stimulation on upper extremity function and cortical perfusion in stroke patients. Clin Neurophysiol 124(10):2008–2015
9. Tanaka E, Iwasaki Y, Saegusa S, Yuge L (2012) Gait and ADL rehabilitation using a whole body motion support type mobile suit evaluated by cerebral activity. In: IEEE international conference on systems, man, and cybernetics, 14–17 Oct 2012
10. http://www.shimadzu.com/an/lifescience/maging/nirs/foire.html, 2013
11. Abibullaev B, An J (2012) Classification of frontal cortex haemodynamic responses during cognitive tasks using wavelet transforms and machine learning algorithm. Med Eng Phys 34:1394–1410
12. Schneiderman JS, Hazlett EA, Chu KW, Zhang J, Goodman CR, Newmark RE, Torosjan Y, Canfield EL, Entis J, Mitropoulou V, Tang CY, Friedman J, Buchsbaum MS (2011) Brodmann area analysis of white matter anisotropy and age in schizophrenia. Schizophr Res 130(1–3):57–67
13. Kleinschmidt A, Obrig H et al (2000) Simultaneous recording of cerebral blood oxygenation changes during human brain activation by magnetic resonance imaging and near-infrared spectroscopy. J Cereb Blood Flow Metab 16:817–826

Chapter 17
Real-Time Monitoring System for Industrial Motion and Optical Micro Vibration Detection

HyungTae Kim and Cheolho Kim

Abstract Sensitive processes in semiconductor manufacturing require the isolation and quick absorption of vibration. In this study, a monitoring system for driving motion and detecting vibration was developed for active vibration isolation and absorption. The monitoring system was an expandable embedded system with functions of motor control, signal conversion, data communication, on-board memory, and IO interfaces. The vibration in this system was at the micro level, so laser gap sensors were used, and a connection was made with a monitoring system using TCP/IP to minimize noise in the data transfer line. A motion and vibration profile was acquired with the developed system.

Keywords Micro vibration • Isolation • Absorption control • Laser gap sensor • Embedded system • Realtime monitoring

17.1 Introduction

The processes of IT core components such as RAM, the CPU, and display require nano-level line widths, and are sensitive to environmental vibration. The vibration in a process has various causes, and must be eliminated quickly. Vibration criteria were proposed for fabrication and precise optics [1]. The performance of an optical system can be improved by just isolating the vibration [2]. Active and passive vibration has been a fairly common issue and many mathematical models can easily be found in previous studies [3]. However, most of these studies have two flaws regarding pay load and vibration detection at the micro level. The weight of the

H. Kim (✉) · C. Kim
Manufacturing System R&D Group, KITECH, 89, YangDae-GiRo, ChungNam, ChenAn 331–825, South Korea
e-mail: htkim@kitech.re.kr
URL: http://www.kitech.re.kr

R. Tutsch et al. (eds.), *Progress in Optomechatronic Technologies*,
Lecture Notes in Electrical Engineering 306, DOI: 10.1007/978-3-319-05711-8_17,

upper structure of an industrial machine reaches tens of tons, and vibration control devices must consider such loads but relevant studies can hardly be found. The heavy weight of the upper structure and low stiffness of vibration control devices result in responses with low frequency less than 10 Hz [4].

Optical devices such as lasers are widely used in vibration measurement. Lasers are the most popular tool for obtaining micro resolution and low-frequency detection. Rygalin et al. [5] developed a laser detector for mechanical resonance in the 1970s. Zastrogin and Zastrogin [6] constructed an optical system to detect vibration using mirrors and splitters without mechanical contact. Laser vibrometers have become popular, and various relevant methods have been reported, one of which involved the Doppler effect [7]. Zhen et al. [8] developed a vibration measurement sensor based on a laser interferometer, which could detect nano-resolution displacement. Eppel et al. [9] applied lasers to detect chattering in a milling process, and showed the possibility of detecting vibration at low frequencies under 50 Hz. Micro and low-frequency vibration is hard to detect, but Ni et al. [10] presented a solution using an MRE core. An isolation and absorption system is being developed for micro vibration using MR fluid and electromagnets for semiconductor manufacturing [12]. An air mount model for micro vibration was also proposed [11], but it was difficult to detect micro vibration. The application of laser sensors to these micro vibration detection will be discussed in this study.

17.2 Monitoring System and Laser Sensors

An air mount is a lower structure to isolate floor vibration and absorb machine vibration. An air mount isolates a heavy surface plate using pneumatic and electromagnetic forces which active vibration control that is adjusted with sensor feedback. The pneumatic forces are provided by an air supply and are transferred to the air mount through valves. A height valve contacts the lower surface of the surface plate and adjusts the open port by vibration. When the position of the height valve becomes higher than the initial position, it regulates the pneumatic pressure to atmospheric pressure. When the position is lower than the initial position, the pneumatic pressure is supplied to the mount. The same principle is applied to the magnetic force. A 650 nm-laser sensor (micro-epsilon ILD-2300) is installed to detect micro and low-frequency vibration by measuring displacement. The displacement is calculated using the laser radiation reflected from the fixture of the surface plate. Due to the high resolution, long data line, and noise in industrial environments, digital data transmission through RJ-45 and TCP/IP was applied. A moving mass was loaded on an XY stage for simulating a virtual machine. High acceleration was a major design factor for the stage, because machine vibration can be generated as in a manufacturing machine. Figure 17.1 shows the structure of the micro vibration control system.

The controller for this system interfaces with the motion, vibration, sampling, and signal in real time, so an expandable embedded controller is essential. As

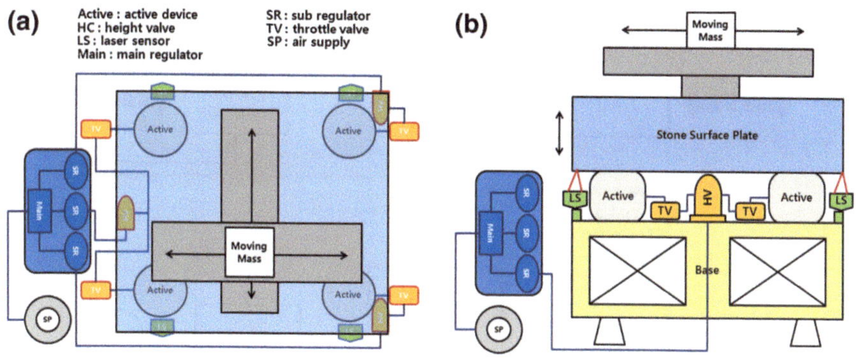

Fig. 17.1 Conceptual structure of micro vibration control system. **a** Top view, **b** side view

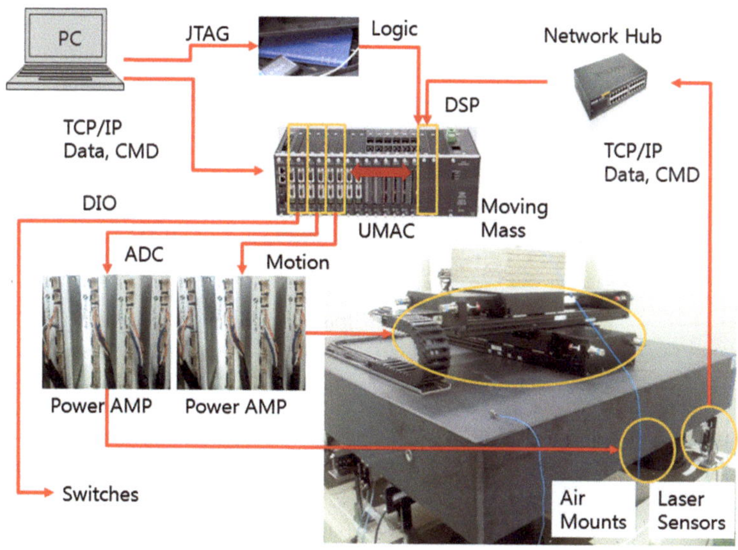

Fig. 17.2 Conceptual diagram for electrical wiring in micro vibration control system

shown in Fig. 17.2, UMAC was used for the main controller and additional DSP boards were developed for decoding data from the laser sensors. The logic of the DSP boards was downloaded through a JTAG emulator. The displacement from the laser sensor was written in the shared memory of the UMAC. An ADC/DAC board was installed for the analog interface and DIO was used for the On-Off or PWM signal. The motion and ADC signals are amplified before driving the motors and the vibration control system.

Multiple lasers and a DSP board were connected by Ethernet. The signal from the sensors must be decoded in real time. The DSP board conducts the signal from the Ethernet port, decodes the data protocol, and transfers the result into the

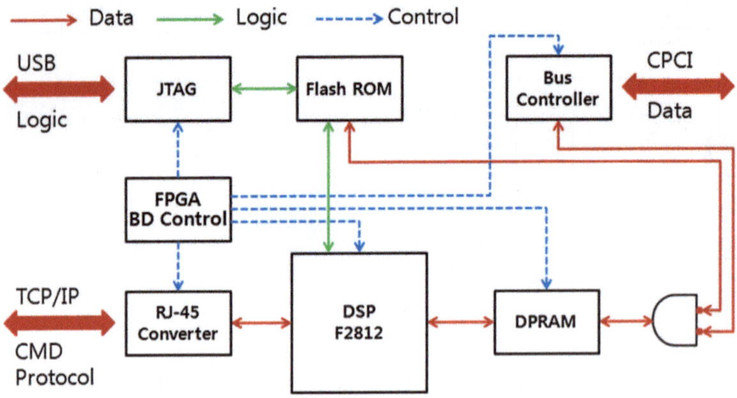

Fig. 17.3 Internal structure of a CPCI DSP board

Table 17.1 Specifications

Device	Contents	Specification
Surface plate	Material	Stone
	Weight	1.5 t
	Size	1.5 m × 1.5 m × 0.25 m
Stage	Stroke	450 mm
	Acceleration	0.5 G
	Accuracy	±12 μm
	Pay load	100 kg
Laser sensor (ILD-2300)	Range	10 μm
	Resolution	±0.15 μm
	Linearity	±2.0 μm
	Measuring rate	49 kHz
Controller (UMAC)	CPU	DSP56303
	Memory	128 k * 24 SRAM
	Axis	4
	Backplane	12 slot CPCI
DSP board (X-com)	CPU	F2812
	Conversion rate	2.2 μs
	Max IP	8

on-board memory of the controller. Figure 17.3 shows the internal structure of the DSP board. The Ethernet control chip decodes and buffers the signal from the LAN port. A DSP chip converts the data in the Ethernet control chip and stores the results in dual port RAM. The data in the dual port RAM are transferred to the on-board memory of the main controller through compact PCI. The DSP logic for the data conversion was constructed using PC development tool and written into flash ROM through the JTAG port. The logic in the flash ROM is loaded into the DSP chip in the booting sequence when the power is on. An FPGA chip was applied to control the main chips on the DSP board. The specifications are shown in Table 17.1.

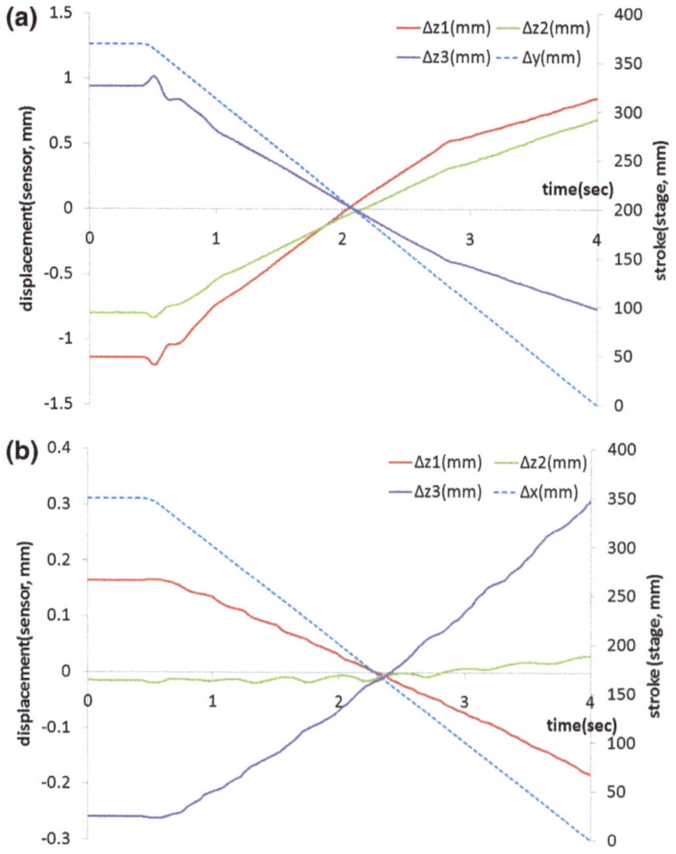

Fig. 17.4 Stage stroke and laser displacement measured in on-board memory. **a** X translation, **b** Y translation

17.3 Experiment and Results

The response of the real-time monitoring was tested using the uniaxial translation of the XY stage. When one of the axes in the stage was driven, the displacements of the laser sensors were stored in the DPRAM of the motion controller. After finishing the translation, the monitoring data in the DPRAM were transferred to a PC for analysis. Figure 17.4 shows the monitoring results during translation. The horizontal axis is time, and the vertical axes are sensor displacement and the actual position of the stage. Figure 17.4a shows the monitoring response of X translation. The relative stroke was 350 mm. Z_1 and Z_2 showed reverse responses to Z_3, because the X location of Z_1 and Z_2 was to the left but Z_3 was to the right. The X translation moved the mass from Z_1 and Z_3, so Z_1 increased but Z_3 decreased. In the case of Y translation, Z_1 and Z_3 showed reverse behavior but Z_2 was constant.

It is considered that the moving mass is placed on the side of Z_1 and Z_3 in the Y direction as shown in Fig. 17.1, so the response of Z_2 was smaller than the other responses. The monitoring data were sampled with 2-ms intervals, and showed that the micro vibration can be monitored with industrial motion in real time. The monitoring system has signal ports and programmable memory to control the active devices. This monitoring system will be upgraded with a control system for vibration isolation and absorption in the future.

17.4 Conclusion

For the purpose of monitoring industrial motion and micro vibration in real time, a DSP board for an embedded controller was developed. The micro vibration was detected by laser sensors, and digital transmission using TCP/IP was applied to minimize noise effects. The DSP board decoded the data from sensors and transferred the decoding result to the on-board memory in the main controller. The response of laser sensors by the uniaxial translation of an XY stage was recorded in the main controller. The monitoring system developed in this study was useful because it can simultaneously monitor motion and micro vibration in real time.

Acknowledgments This work was funded and supported by the Korea Institute of Industrial Technology and Ministry of Trade, Industry and Energy. The authors are grateful to Delta Tau Korea (http://www.deltatau.co.kr) for technical support, and to RMS Technology (http://www.rmstech.co.kr) for air mounts.

References

1. Bessason B, Madshus C, Frøystein HA, Kolbjørnsen H (1999) Vibration criteria for metrology laboratories. Meas Sci Technol 10(11):1009–1014
2. Heo JW, Chung JT (2002) Vibration and noise reduction of an optical disk drive by using a vibration absorber. IEEE Trans Consum Electron 48(4):874–878
3. Kato T, Kawashima K, Sawamoto K, Kagawa T (2010) A new, high precision, quick response pressure regulator for active control of pneumatic vibration isolation tables. Precision Eng 34:43–48
4. Kim HT, Kim CH, Kang SB, Moon SJ, Lee GS (2013) A hybrid structure of dual stators and a pneumatic spring for resonance control in an air mount. J Electromagn Anal Appl 5(3):114–119
5. Rygalin VG, Grechinskii DA, Sobolev AB (1975) Use of laser equipment to detect mechanical resonances and to stabilize vibration parameters at a specified value. Meas Tech 18(5):722–723
6. Zastrogin YF, Zastrogin OY (1992) Methods and devices for vibration control. Meas Tech 35(1):89–94
7. Kim JK, Chang TH, Kim HS, Kang MS (2004) Adaptive compensation of motional disturbances for laser Doppler vibrometer. In: Proceeding of the IEEE Industrial Electronics Society, Piscataway, NJ, 3:2081–2085

8. Zhen S, Chen B, Yuan L, Li M, Liang J, Yu B (2010) A novel interferometric vibration measurement sensor with quadrature detection based on 18 wave plate. Opt Laser Technol 42:362–365
9. Eppel A, Enikov ET, Insperger T, Gabor S (2010) Feasibility study of optical detection of chatter vibration during milling. Int J Optomechatronics 4:195–214
10. Ni YQ, Ying ZG, Chen ZH (2011) Micro-vibration suppression of equipment supported on a floor incorporating magneto-rheological elastomercore. J Sound Vib 330:4369–4383
11. Kim HT, Kim CH, Kang SB, Lee KW, Baek JH, Han HH (2012) A 3DOF model for an electromagnetic air mount. Adv Acoust Vib 2012:218429
12. Kim HT, Kim CH, Choi SB, Moon SJ and Song WG (2012) Triple mechanism of an active air spring combining pneumatics, electro-magnets and MR fluid in an air mount. In: Proceedings of international conference on sound and vibration

Chapter 18
Thermal Manipulation Utilizing Micro-cantilever Probe in Scanning Electron Microscopy

Anthony Yu-Tung Wang, Cheng-Chun Huang, Yao-Chuan Tsai, Ming-Dao Wu, Dao Liang, Po-Jen Shih and Wen-Pin Shih

Abstract We present a novel design for a sensitive temperature micro-probe, situated at the tip of an atomic force microscopy cantilever. The temperature-sensing element utilizes a platinum resistance thermometer, which is well-known for its measurement reproducibility and chemical inertness. The probe is fabricated using conventional clean room techniques and then processed by focused ion beam milling and material deposition. The probe is able to be mounted to a movable platform inside a scanning electron microscope, enabling simultaneous characterization of a sample's surface temperature and material topology. Furthermore, by reversing the detection mechanism of the platinum resistance thermometer, localized sample surface heating can be achieved, allowing small-scale sample manipulation and characterization. Such an ability to simultaneously characterize a material's surface topology and temperature has not been previously reported in literature, and lends great practicality in the fields of materials research and integrated circuits diagnostics.

Keywords Micro-cantilever • Thermometer • Focused ion beam • Atomic force microscopy

A. Y.-T. Wang
University of Waterloo, Waterloo, ON N2L 3G1, Canada
e-mail: aytwang@uwaterloo.ca

A. Y.-T. Wang · C.-C. Huang · Y.-C. Tsai · M.-D. Wu · D. Liang · W.-P. Shih (✉)
National Taiwan University, Taipei City 10617, Taiwan
e-mail: wpshih@ntu.edu.tw

P.-J. Shih
National University of Kaohsiung, Kaohsiung City 81148, Taiwan

R. Tutsch et al. (eds.), *Progress in Optomechatronic Technologies*,
Lecture Notes in Electrical Engineering 306, DOI: 10.1007/978-3-319-05711-8_18,

18.1 Introduction

Moore's law states that the total number of transistors on integrated circuits will double every 18 months [1]. Ever since its introduction in the year 1965, electronic, optoelectronic devices and related circuits have followed the described trend infallibly.

Modern integrated circuits involve a complex myriad of transistors, interconnects, and layers of circuitry. The drastic miniaturization of such circuits and increase in transistor switching speeds have contributed to localized heating problems [2–4]. Such problems may be caused by differences in power dissipation, materials, geometries, or manufacturing methods, and can cause problems for today's sensitive circuits. Common heat-related issues include inadequate diffusion of silicon dopants [5]; unwanted changes in electron or hole carrier mobility [6]; and the triggering and propagation of lattice dislocations. If severe enough, these problems can cause circuit failure by open- or short-circuit, and must be avoided [7].

Lattice dislocations triggered by local thermal stress could propagate to the surface of the sample, thereby allowing surface scanning methods to give insight into the failure modes of these systems. With these in mind, it is apparent that there is a need to develop a novel instrument capable of simultaneously characterizing the temperature distribution and surface topology of a given sample. Currently, no such instrument exists, and the development of this technology will aid greatly in the performance, reliability, and failure analyses of integrated circuits and related systems [8].

18.1.1 Atomic Force Microscopy as a Basis for the Thermal Probe

The atomic force microscope ('AFM') was invented by Binnig et al. [9] in the year 1986. The AFM is widely used in academia and industry as a valuable research tool for the imaging, measuring, and manipulation of matter at the nanoscale. Conventional AFM setups are capable of horizontal spatial resolutions in the tens of nanometres and vertical resolutions in the order of fractions of a nanometer [9, 10].

Generally, the AFM probe is connected to piezoelectric actuators that move to scan the needle across the sample surface. The AFM cantilever deflects in response to repulsion or attraction forces between its tip and the sample surface, generating a vertical map of the surface topology. The probe tip is very sharp (radii of tips are typically below 10 nm) and is usually in a rectangular pyramidal shape. The probe is situated at the long end of a cantilever beam, which magnifies the up-down deflections of the AFM probe while it traces the sample surface contours. The probe's vertical deflection is measured by a laser interferometer focused on the reverse side of the cantilever. The backside of the cantilever is usually coated with a reflective metal such as aluminum to increase the reflectivity.

The cantilever deflection is then used to calculate force and the data is then used as a feedback mechanism for the probe operation [10].

Apart from the dynamic-mode AFM mentioned above, several other AFM technologies currently exist. Two notable examples include the field emission and microfluidic AFM. These technologies differ in their operating principles from the dynamic AFM. The field emission AFM operates based on the physical phenomenon of field emission between the probe tip and the sample surface. A drawback of this technology is that it requires the sample to be conductive and the measurement technique is plagued by high noise levels caused by field emission current instabilities [11]. The microfluidic AFM conducts measurements in a liquid environment, housed in a polymeric chamber. It is useful for ex vivo characterization of water-soluble substances, such as proteins, DNA, and other biomolecules [12]. However, the surface image is substantially blurred at moderate-to-high contact pressure, as the polymeric chamber easily deforms under the contact stress [13].

To this day, the atomic force microscope as well as its technological ancestor, the scanning tunneling microscope, remain among the most important materials characterization technologies used in research throughout the world. However, despite the large variety of techniques available, so far there are none that are able to simultaneously characterize the surface topology and temperature distribution of a given sample. The ability to perform these tasks using a contact-based method is a crucial improvement to the currently available methods. The information gathered using such an instrument can aid tremendously in integrated circuits diagnostics and materials analysis.

Since the current technology of the atomic force microscope is already very sophisticated, it is reasonable to base the design of the combination sensor on the existing AFM system. The AFM probe will be modified to record thermal information by attaching a temperature sensor to its tip. The probe will not interfere with the AFM's original surface characterization and physical sample manipulation capabilities. Furthermore, by applying a controlled current through the thermometer, we can achieve a heating effect, thereby allowing another method of precise sample surface manipulation. The following section will discuss the selection process for an ideal material to be used as the temperature sensor, and review the various benefits it will give over conventional methods of small-scale thermal characterization.

18.1.2 Conventional Thermal Characterization Methods

Currently, there is a large variety of methods available for the measurement of temperature distribution on small sample surfaces. These usually rely on the dependence of the material surface's physical properties on temperature to determine the surface temperature. These techniques have varying thermal and spatial resolutions as well as acquisition time, and are often limited in their use depending on the sample material, structure, and topology [8, 14].

Coating measurements such as liquid crystal thermography allow a facile way to estimate temperature using the naked eye, but requires an exposed device surface for best resolution, suffers from limited temperature range, and risks the damaging of sensitive electronics. Mechanical contact methods such as using a micro-thermocouple exhibit good temperature resolution, at the cost of slow detection speed and poor spatial resolution. Optical methods exhibit very good lateral resolution and measurement speed, but lack the ability to characterize the sample surface. Lastly, direct sensor integration onto the devices involves more effort in the circuit design, and any significant changes in the device's circuitry will render the sensor integration unusable [14–17].

These methods are all capable of thermal characterization on the small-scale. However, so far there are no methods that can combine speed, sensitivity, range, and ease of use with the ability of sample surface imaging. Thus, under the increasing pressure of having to characterize shrinking ICs, a major motivation for the development of a small-scale temperature/topology sensor was created.

18.1.3 The Platinum Resistance Thermometer as an Ideal Temperature Sensor

The platinum resistance thermometer ('PRT') is widely used in research and industry as a reliable way to measure temperature in almost all environments. They function based on platinum's changing electrical resistance in response to different temperatures. By measuring the resistance of the platinum wire and comparing it to a calibrated standard, the temperature can be accurately calculated. The PRT is internationally recognized by the IEC 60751 and ASTM E-1137 standards [18, 19], and offer many excellent properties: high sensitivity, response linearity, reproducibility, material stability, and detectable temperature range. Amongst other common resistance thermometer metals (Ni, Cu, and others), platinum's resistance-temperature response curve exhibits the most linearity over the largest temperature range [20, 21]. Copper metal was also considered as a material for the construction of the resistive thermometer. However, because of copper's low thermal oxidation temperature and susceptibility to chemical corrosion, it was rejected in favour of platinum [22].

All of the mentioned benefits make the PRT the ideal temperature sensing element for a combination temperature/topology probe. Thus, by modifying an existing AFM probe with a PRT, the goal of fabricating an instrument that can simultaneously and reliably measure a sample's topology and temperature distribution should be achievable.

18.2 Design and Fabrication of the PRT-AFM Probe

Numerous designs for the probe were proposed and scrutinized for their functionality and ease of fabrication. The final design of the probe required two electrodes that are situated at the tip of the AFM cantilever. The two electrodes

Fig. 18.1 Prototype design of the PRT-AFM probe

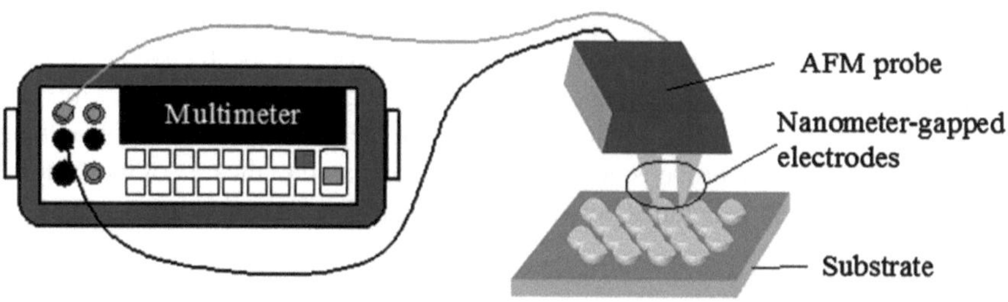

Fig. 18.2 The complete PRT-AFM system

are electrically separated and are connected to separate contact pads on the probe body. An illustration of the completed probe can be seen in Fig. 18.1. Note that the fabricated probe will from hereon be referred to as 'PRT-AFM'.

A platinum nanowire bridges the gap between the two electrodes on the probe tip, forming the main temperature-sensing element of the temperature-topology sensor. The contact pads on the probe are conductive and are used later on for electrical resistance measurements. They are patterned by photoresist lift-off process.

In the completed system, the probe will have its contact pads wire-bonded to an external circuit board. The probe will come into contact with the sample surface, as shown in Fig. 18.2. The resistance of the platinum nanowire will be measured with a digital multimeter, and using a previously-calibrated resistance standard curve, the temperature at the probe tip can be calculated.

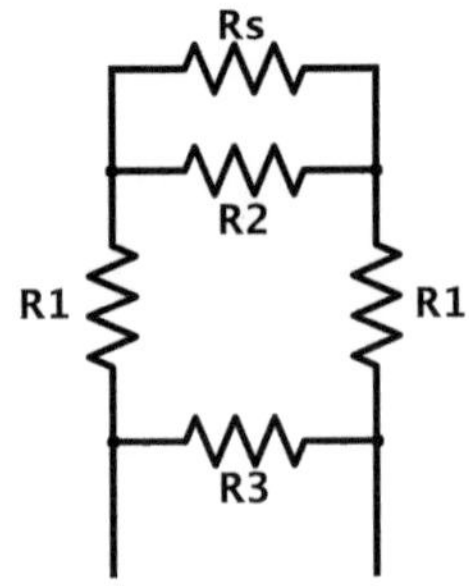

Fig. 18.3 Circuit model for the completed PRT-AFM

18.2.1 Leakage Current and Circuit Analysis

The PRT-AFM probe was designed in such as way that the net measured change in the resistance of the probe at the contact pads is the change in resistance observed by the platinum resistance thermometer at the tip of the probe. A simple DC circuit analysis (Fig. 18.3) will demonstrate this.

The resistance models in the figure are as follows: R_1 is the resistance of the deposited platinum lead wire from the contact pads to the temperature-sensing platinum wire (R_s), and R_2 and R_3 represent the resistance between the separate electrodes on the cantilever and body side of the AFM probe, respectively. These finite resistances are the result of leakage current occurring between the lead wires and metal contacts, and should be considered. Note that the resistance of the gold contact pads on the body of the probe is negligible when compared to the rest of the system.

For the equivalent resistance between R_s, R_2, and R_1, we can write

$$R_{1,2,s} = R_p + 2R_1, \tag{18.1}$$

where $R_p = \frac{R_s \cdot R_2}{R_s + R_2}$. The total equivalent resistance for the probe is then

$$R_t = \frac{(R_p + 2R_1)R_3}{R_p + 2R_1 + R_3}. \tag{18.2}$$

We note that R_3 is large compared to the rest of the system, since R_3 represents the resistance between the two gold contacts separated by at least 140 μm on the non-conductive surface of the probe. The total resistance then becomes:

$$R_t \approx R_p + 2R_1 = \frac{R_s \cdot R_2}{R_s + R_2} + 2R_1. \tag{18.3}$$

We further note that R_2 is large, since it represents the resistance between the two electrodes at the cantilever tip, separated by an air gap of about 100 nm (from SEM results). In other words,

$$R_t \approx R_s + 2R_1. \tag{18.4}$$

Finally, we note that because resistance of a wire is given by the formula $R = \frac{\rho \cdot l}{A}$ and that the diameter of the temperature sensing wire is smaller than that

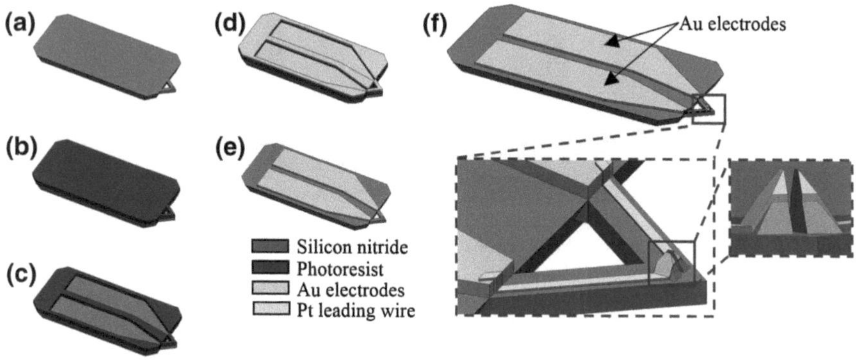

Fig. 18.4 Fabrication process of the PRT-AFM, involving **a–c** photoresistive patterning of contact pads, **d** deposition, **e** lift-off of gold contact pads, and **f** deposition of resistive platinum nanowire

of the lead wires, $R_s \gg R_1$. The measurable resistance is then simply the change in resistance of the sensing wire:

$$R_t \approx R_s. \tag{18.5}$$

18.2.2 Fabrication of the PRT-AFM

The fabrication of the PRT-AFM involved the deposition and patterning of chromium and gold layers on the probes for electrical contact, focused ion beam ('FIB') splitting of the probe tip to form the two electrodes, and deposition of the platinum nanowire to form the temperature-sensing elements. A diagram of the fabrication process can be seen in Fig. 18.4.

Using ion beam-induced deposition ('IBID'), we deposited a platinum nanowire between the two electrodes. The nanowire would bridge the air gap between the two fabricated electrodes, and is the major thermal sensing element of the completed PRT-AFM. However, problems were encountered during deposition of the nanowire. While the IBID process is well-implemented, the 3-dimensional positioning and geometry of the AFM probe inside the FIB chamber, in relation to the incoming ion beam, was very difficult. The AFM probe and its two electrodes must be positioned in such a way that allows the ion beam to deposit the platinum nanowire between the electrodes. This is most convenient when the platinum nanowire is being deposited perpendicularly onto the electrode surface. Since the nanowire is growing vertically from the deposition surface, ion scattering effects at the wire tip were of crucial consideration in the deposition process [23]. The deposited platinum nanowire also exhibited surface roughness as a result of impurities from the organic precursor gas and re-deposition of gallium metal from the ion beam [24, 25].

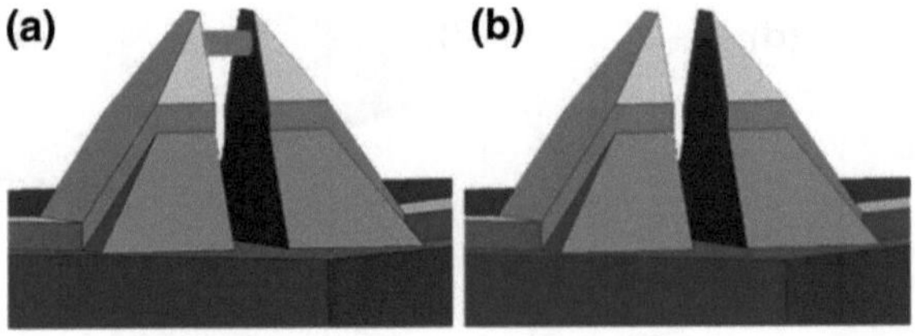

Fig. 18.5 Comparison of PRT-AFM probe tip designs **a** with and **b** without bridging nanowire

Furthermore, after each adjustment of the sample stage in the FIB, the electron and ion beams must be manually aligned again. Since the ion beam physically impels and erodes on the sample surface, turning it on for an extended period of time—such as required during electron and ion beam alignments—is not good for the surface of the sensitive AFM probe.

After a few failed fabrication attempts, a change was made to the prototype design that would produce a better yield. The original design featured a platinum nanowire bridging the air gap between the two electrodes, forming the major thermal sensing element. However, fabrication difficulties as mentioned above reduced the successful yield greatly. Therefore, a new design (Fig. 18.5) was proposed that eliminated the nanowire between the two electrodes.

Instead of the platinum nanowire bridging the two separate electrodes (Fig. 18.5a), the wires leading to the two electrodes will be used as the thermal sensing elements (Fig. 18.5b). Under this new design, there will be an air gap between the two measurement electrodes. However, this does not prevent the electrical measurement of the platinum lead wires' resistance because the gap is in the order of nanometres. Current leakage will occur between the two electrodes, and is enough to allow resistance measurement.

18.3 Detailed Fabrication Methods

AFM probes of the 'PNP-DB' variety were obtained from NanoWorld (Neuchâtel, Switzerland), and are made of silicon nitride cantilever on an insulated Pyrex body.

The AFM probes were fixed to a silicon wafer with SPR220-7 photoresist (Stanford University, Stanford, CA). The photoresist was exposed using a MA6 mask aligner (Karl-Süss, Garching, Germany) for 75 using 365.4 nm I-line, and then baked at 90 °C for 30 min. The photoresist was then developed, leaving the exposed patterns for the deposition of chromium and gold layers.

Deposition of chromium and gold layers was done using EBPVD (F.S.E. Corporation, New Taipei City, Taiwan) in a vacuum of 6.78×10^{-8} mmHg. Film thickness and deposition rate were measured in situ using a CRTM-6000 quartz crystal microbalance (ULVAC, Methuen, MA). Electron emission intensity was adjusted using a HPS-510S power supply (ULVAC).

To promote atom-to-atom adhesion, the film growth rate was limited to 0.2 Å s^{-1} for the first and last 10 Å of deposited film thickness, and then between 0.60 to 0.8 Å s^{-1} for the rest of the deposition. Chamber temperature was kept

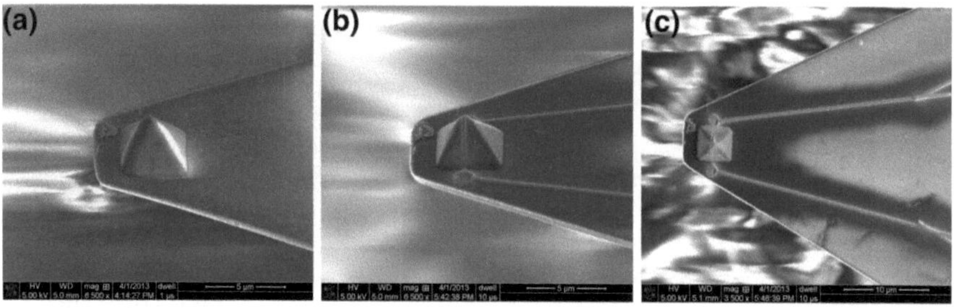

Fig. 18.6 Comparison of unmodified (**a**) and modified (**b**) PRT-AFM probes using FIB. **c** *Top* view of probe

below 80 °C to avoid thermal damage to the photoresist. Total deposited film thickness was 164 Å for the chromium layer and 160 nm for the gold layer.

After deposition, the probes were immersed in acetone for up to 1 h to lift-off the unwanted metal patterns. In the case where the resist is not completely dissolved away in the bath, manual brushing using a fine brush was performed.

Using a Nova 600 NanoLab dual-beam SEM/FIB (FEI Company, Hillsboro, OR), the AFM cantilever tip was first split, then platinum wire deposited to form the temperature sensor and lead wires. Electron acceleration voltage was set at 5 kV with a current of 98 pA, and the ion acceleration voltage was set at 30 kV with current of 93 pA. These parameters were chosen for the best balance between fabrication resolution and speed.

The AFM probe tip was split with a Ga^+ beam using preset protocols in the FIB software. The milling was automated by the software following a silicon etch protocol. Similarly, the IBID of platinum nanowires was completed using preset software protocols with the precursor gas trimethylcyclopentadienyl-platinum $(CH_3)_3Pt(CpCH_3)$. The temperature-sensing platinum nanowire was deposited first, followed by the lead wires that connected the PRT to the gold contact pads on the AFM body. A scanning electron micrograph of the cantilever tip before and after modification can be seen in Fig. 18.6.

After FIB fabrication, the probes wire-bonded to an external circuit board. The resistances between the contact pads on the AFM probe and the circuit board were measured after wire-bonding to ensure a good electrical connection.

18.4 Results and Discussion

The completed PRT-AFM probe was inspected under optical and electron microscopes to verify its visual appearance (see Figs. 18.6 and 18.7).

Note the visible platinum nanowires under the magnified view of the image. SEM measurements show that the gap between the two electrodes at the cantilever tip was around 120 nm, and that the width of the temperature-sensing platinum

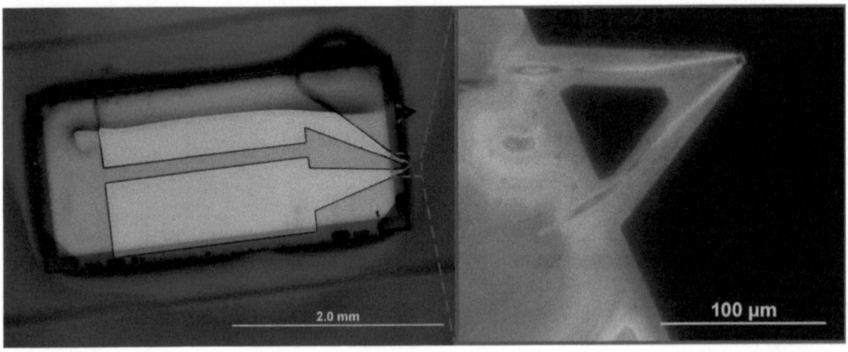

Fig. 18.7 Optical microscopy photograph of the completed PRT-AFM

nanowire was less than 100 nm. These dimensions are desirable for nanoscale temperature and topology characterization, and are within the design specifications of our probe. However, there are opportunities for further improvement. For example, the gap size between the electrodes could be reduced by increasing the acceleration voltage of the FIB (thereby reducing the ion spot size). However, we note that increasing the ion velocity will also cause unwanted effects such as probe surface damage and the delamination of the chromium and gold layers. Therefore, an optimum set of parameters must be found that minimizes both the ion beam spot size and probe surface damage.

18.5 Conclusions

A design for a sensor that can simultaneously characterize a given sample's temperature and surface topology was proposed. We have shown positive results in the probe's fabrication. The design was fabricated on an atomic force microscopy probe modified using FIB and other common integrated circuit fabrication techniques. Optical and electron microscopy were used to study every step of the fabrication process.

We conclude that FIB can be successfully used to modify the AFM probe into the desired electrodes by ion milling. The gap between the two electrically-insulated probes after the FIB modification is also of a desired width, in the order of hundreds of nanometres. This allows for the probe to operate at the nanoscale, and matches the original design requirement.

Furthermore, the deposition of platinum resistance nanowire to be used as the temperature-sensing element in the PRT-AFM probe can be accomplished with the FIB using standard platinum deposition techniques. The deposited nanowire was of a desired width, and met the design requirements. The growth of the bridging nanowire in 3-dimensional space between the two electrodes, while theoretically possible, is currently unachievable. The authors are working on a solution to this problem.

Acknowledgments The author would like to thank the co-operative education program of the University of Waterloo, Canada for the opportunity to conduct research abroad at the National Taiwan University. The author would also like to thank the Tiny Machines and Mechanics Laboratory and the Nano-Electro-Mechanical-Systems Research Center of National Taiwan University for hosting this project, and providing funding and the necessary facilities to carry out the research. This project was supported by the National Science Council of Taiwan, under contract No. 102-2221-E-002-242.

References

1. Moore GE (1965) Cramming more components into integrated circuits. Electronics 38:114–117
2. Liu TJK (2012) Bulk CMOS scaling to the end of the roadmap. Presentation (June 2012) Symposium on VLSI Circuits
3. Rotem E, Hermerding J, Aviad C, Harel C (2007) Temperature measurement in the Intel Core^{TM} Duo processor. Technical report, Intel Corporation (2007)
4. De Wolf I (1996) Micro-Raman spectroscopy to study local mechanical stress in silicon integrated circuits. Semicond Sci Technol 11:139–154
5. Park H, Jones KS, Slinkman JA, Law ME (1993) The effects of strain on dopant diffusion in silicon. In: Technical digest, international electronic devices meeting, international electronic devices meeting (IEDM), IEEE Conference Publications (December 1993)
6. Degraeve R, De Wolf I, Groeseneken G, Maes H (1994) Analysis of externally imposed mechanical stress effects on the hot-carrier-induced degradation of MOSFET's. In: 32nd annual reliability physics symposium, IEEE, 32nd annual proceedings, IEEE international
7. Hu SM (1991) Stress-related problems in silicon technology. J Appl Phys 70:R53–R79
8. Christofferson J, Maize K, Ezzahri Y, Shabani J, Wang X, Baskin AS (2007) Microscale and nanoscale thermal characterization techniques. In: International conference on thermal issues in emerging technologies: theory and application—THETA
9. Binnig G, Quate CF (1986) Atomic force microscope. Phys Rev Lett 56:930–933
10. Kaupp G (2006) Atomic force microscopy, scanning nearfield optical microscopy and nanoscratching. Springer, Berlin, pp 1–86
11. le Fébre AJ, Abelmann L, Lodder JC (2008) Field emission at nanometer distances for high-resolution positioning. J Vac Sci Technol B 26:724–729
12. Sulchek TA, Qiu SR, Noga DJ, Schoenwald DK (2012) Molded microfluidic fluid cell for atomic force microscopy . https://www.google.com/patents/US8214917
13. Schoenwald K, Peng ZC, Noga D, Qiu SR, Sulchek T (2010) Integration of atomic force microscopy and a microfluidic liquid cell for aqueous imaging and force spectroscopy. Rev Sci Instrum 81:053704
14. Cutolo A (1998) Selected contactless optoelectronic measurements for electronic applications (invited). Rev Sci Instrum 69:337–360
15. Altet J, Dilhaire S, Volz S, Rampnoux JM, Rubio A, Grauby S, Lopez LDP, Claeys W, Saulnier JB (2002) Four different approaches for the measurement of IC surface temperature: application to thermal testing. Microelectron J 33:689–696
16. Altet J (2006) Dynamic surface temperature measurements in ICs. Proc IEEE 94:1519–1533
17. Kölzer J, Oesterschulze E, Deboy G (1996) Thermal imaging and measurement techniques for electronic materials and devices. Microelectron Eng 31:251–270
18. International Electrotechnical Commission: IEC 60751: Industrial platinum resistance thermometers and platinum temperature sensors
19. ASTM International: ASTM E1137: Standard specification for industrial platinum resistance thermometers

20. Yamazawa K, Anso K, Widiatmo JV, Tamba J, Arai M (2011) Evaluation of small-sized platinum resistance thermometers with ITS-90 characteristics. Int J Thermophys 32:2397–2408
21. Yang I, Song CH, Gam KS, Kim YG (2012) Long-term stability of standard platinum resistance thermometers in a range between 0.01 °C and 419.527 °C. Metrologia 49:803–808
22. Langlands RC (1964) A stable copper resistance thermometer for field use. J Sci Instrum 41:478
23. Bret T, Utke I, Hoffmann P, Abourida M, Doppelt P (2006) Electron range effects in focused electron beam induced deposition of 3D nanostructures. Microelectron Eng 83:1482–1486
24. Tripathi SK, Shukla N, Kulkarni VN (2008) Correlation between ion beam parameters and physical characteristics of nanostructures fabricated by focused ion beam. Nucl Instrum Methods Phys Res B 266:1468–1474
25. Tripathi SK, Kulkarni VN (2009) Evolution of surface morphology of nano and micro structures during focused ion beam induced growth. Nucl Instrum Methods Phys Res B 267:1381–3385

Chapter 19
Sound-Source Tracking and Obstacle Avoidance System for the Mobile Robot

Bo-Yeon Hwang, Sook-Hee Park, Jong-Ho Han, Min-Gyu Kim and Jang-Myung Lee

Abstract In this paper, the study for the obstacle avoidance of the sound-source tracking mobile robot is introduced. By using the cross-correlation, it is possible to estimate the time difference for sound waves reaching more than two sound-source obtaining devices. Also, through the sound-source obtaining devices and carrying out the geometric interpretation, the location of each sound-source can be found. Based on the suggested application of the motion algorithm, when the mobile robot is in motion, it creates the virtual repulsive force with any obstacle, while applying the virtual attraction force with the reference point in order to avoid obstacles more smoothly and effectively.

Keywords Microphone array • Tracking • Obstacle avoidance • Sound source

19.1 Introduction

The technology for recognizing the location or direction of a sound-source by using a microphone, and the one for recognizing human voices have been developed greatly until now. The technique for detecting the location and direction of

B.-Y. Hwang · S.-H. Park · J.-H. Han · M.-G. Kim · J.-M. Lee (✉)
Department of Electrical Engineering, Pusan National University, Busan, South Korea
e-mail: jmlee@pusan.ac.kr

B.-Y. Hwang
e-mail: boyeon1696@pusan.ac.kr

S.-H. Park
e-mail: sookhee7379@pusan.ac.kr

J.-H. Han
e-mail: gkswhdgh114@pusan.ac.kr

M.-G. Kim
e-mail: mingyu1696@pusan.ac.kr

R. Tutsch et al. (eds.), *Progress in Optomechatronic Technologies*,
Lecture Notes in Electrical Engineering 306, DOI: 10.1007/978-3-319-05711-8_19,

a sound-source is called as the location recognition of a sound-source, while the one for analyzing and recognizing human voices is called as the voice recognition. There are various techniques used in this study. In order to use the sound-source, it is required to have at least more than two sound-source obtaining devices or microphones. The sound signals reach each microphone with TDOA (Time Difference of Arrival) and DSA (Difference of Sound Amplitude). Based on TDOA, the methods of estimating time differences mainly include the one of measuring specific values and the one of measuring the level of similarity for the signals. The method of estimating the maximum or threshold values could have the advantage of being fast, but it shows the disadvantage of being weak to noises. Meanwhile, it is possible to use the method the method of cross-correlation or generalized cross-correlation (the application of PHAT) and the difference method which are strong against noises [1–3]. Even if the difference method could cause a low level of accuracy compared to the method of cross-correlation when the types of waveform for two signals are different, it is still useful in case of spatio-temporal limitations due to the relatively low level of arithmetic operation. Recently, researchers have studied the method of estimating the location of a sound-source after mapping the cross-correlation functions on the space coordinates by using proper mapping functions without estimating the time delays through the cross-correlation functions [4]. Also, such others methods as HRTF (Head Related Transfer Function), which is used as the sound-source system of a robot, the beamforrming method, and the one of using artificial ears which imitate those of human beings have been studied [5–7].

In order for a mobile robot to move, it is necessary to carry out the mechanical interpretation of the robot and recognize its location based on the interpretation before planning routes. One of the representative methods of recognizing the location of a mobile robot is the dead-reckoning method which uses the encoded data of a motor [8, 9]. Recently, the study for the recognition of locations based on a sensor has been actively carried out. In the actual environment in which a mobile robot moves, there are obstacles. Therefore, it is necessary to study the algorithm of recognizing obstacles in the surrounding environment and avoiding them before reaching the reference point [4]

PFM (Potential Field Method) is the method applied based on the assumption that there is the attraction force between a mobile robot and its reference point and there is the repulsive force against obstacles [10, 11]. Such a method is related to the virtual impedance algorithm which controls locations and maintains a certain amount of force based on the feedback force by modeling the interaction between an uncertain environment and a robot [12].

In this paper, the experiment for recognizing the locations of moving sound-sources by using the difference method based on TDOA is carried out. In case of a robot meeting obstacles while moving along the route for detecting sound-sources, the simulation for avoiding the obstacles is executed by applying the virtual impedance technique.

First of all, in Chap. 2, the algorithm of recognizing the locations of sound-sources is described. In Chap. 3, the virtual impedance method, which is the algorithm of avoiding obstacles suggested in this paper, is explained. In Chap. 4, the results given by the experiment and the simulations are discussed. In the last chapter, a conclusion is suggested to finish this paper.

19.2 Algorithm of Recognizing the Locations of Sound-Sources

When a sound is generated from one point sound-source, the sound waves are transferred to the sound-source obtaining device at a constant speed. Therefore, the signals reaching more than two sound-source obtaining devices could have time differences. Since the time differences can vary based on the locations of sound-sources, they can be used to estimate such locations. In this paper, the difference method is used to estimate the time differences for the arrival of the sound waves reaching three sound-source obtaining devices. Also, the locations of the sound-sources are measured by using the sound-source obtaining devices arranged in a line and the geometric interpretation with the sound-sources.

19.2.1 Arrival Time Difference

In order to recognize locations by using sounds, it is required to define the sound signals received by each microphone first. When the sound-source signals sent from one point sound-source are received by the receiving part consisting of three microphones, the signal sound $x_i(t)$, which is received by the ith microphone for certain time t, can be expressed as follows.

$$x_i(t) = s(t + t_i) + n_i(t) \qquad (i = 1, 2, 3) \tag{19.1}$$

In the above equation, $s(t)$ is the signal generated from a sound-source, $n_i(t)$ is the external noise received by a microphone, and t_i is the time required for a signal to be received by the ith microphone. Based on the first microphone, the time difference for a signal to be received by the ith microphone is Δt_i and can be expressed as follows.

$$\Delta t_i = t_i - t_1 \qquad (i = 2, 3) \tag{19.2}$$

Also, R_i, which is the distance between a sound-source and the ith microphone, can be expressed in the following equation in terms of t_i.

$$R_i[\mathrm{m}] = t_i[\mathrm{s}] \cdot c[\mathrm{m/s}] \qquad (i = 1, 2, 3) \tag{19.3}$$

Here, c represents the speed for the arrival of sound waves. The required speed can vary based on the substances delivering sounds. In the atmosphere with the room temperature of 25 °C, the speed is estimated to be about 340 m/s.

19.2.2 Difference Method

The difference method can be used to estimate the time differences between two estimated signals. The difference method is the one of estimating the level of similarity by adding the differences between two signals. When the sum of the differences between two signals becomes smaller, the level of similarity tends to become greater. When one of the two signals is moved by Δt_{12}, Δt_{12} shows the time when the level of similarity becomes the greatest and represents the time difference for the arrival of two signals.

The estimated two wave forms x_1 and x_2 can be expressed in the n number of discrete data groups as follows.

$$\begin{cases} x_1(0),\ x_1(1),\ \ldots,\ x_1(n-1) \\ x_2(0),\ x_2(1),\ \ldots,\ x_2(n-1) \end{cases} \tag{19.4}$$

Here, the equation of estimating $S_{x_1,\,x_2}$by adding the differences of the two signals can be expressed as follows.

$$S_{x_1,\,x_2} = \sum_{k=0}^{n-1} \sqrt{(x_1(k) - x_2(k))^2} \tag{19.5}$$

When the second signal is moved by j_M, the difference coefficients can be expressed in the functions for j_M as shown in Eq. (19.6).
j_M contains the values between 0 and n − 1. By multiplying j_M having the greatest level of similarity with the sampling time, it is possible to estimate Δt_{12} which is

$$S_{x_1,\,x_2}(j_M) = \sum_{k=0}^{n-1} \sqrt{(x_1(k) - x_2(k + j_M))^2} \tag{19.6}$$

the time difference for the arrival of the two signals. The following equation shows such a process in a mathematical form.

$$\Delta t_{12} = \frac{\arg\min}{j_M = 0,\ 1,\ \ldots,\ n-1} \left| S_{x_1,x_2}(j_M) \right| \times t_{sampling} \tag{19.7}$$

19.2.3 Algorithm of Recognizing the Locations of Sound-Sources

Based on the algorithm of analyzing the sound-sources executed before, it is possible to estimate the distance and angle of each sound-source. By arranging one sound-source and three microphones, M_1, M_2 and M_3 on the same plane with a constant distance ($l_1 = l_2 = 45$ cm), it is possible to estimate R_1, R_2 and R_3, which show the distances between the detected sound-source and each microphone, and the angle θ.

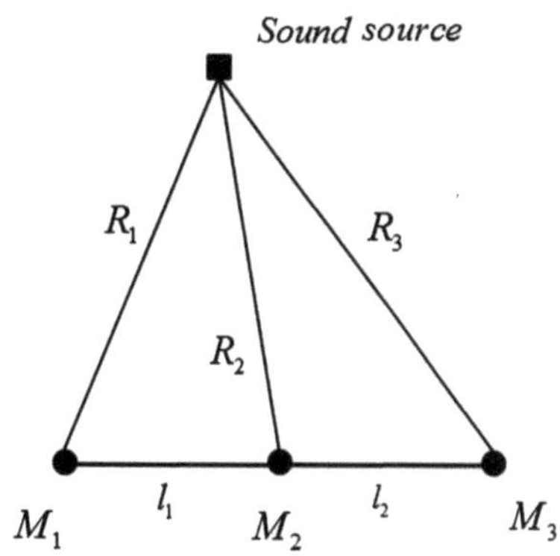

Fig. 19.1 The geometric structures of the sound-source and each microphone in the two-dimensional space

Estimation of the Distance in the Two-dimensional Space.

In Fig. 19.1, R_1, R_2 and R_3 show the distances between the detected sound-source and each microphone. If R_2, which is the distance with M_2 located in the center, is shown as d, which is the distance between a mobile robot and a sound, R_1, R_2 and R_3 can be estimated as follows.

$$\begin{cases} R_2 = c \cdot t_2 = d \\ R_1 = c \cdot t_1 = d + c \cdot \Delta t_{12} \\ R_3 = c \cdot t_3 = d + c \cdot \Delta t_{23} \end{cases} \tag{19.8}$$

In Eq. (19.8), c is the speed of the sound wave (340 m/s), while t_i $(i = 1, 2, 3)$ shows the time value required for the arrival of the estimated signal sound at the ith microphone. Δt_{12} and Δt_{23} represent the arrival time differences estimated through the difference method mentioned before. By using Heron's formula, the sizes of the two triangles ΔSM_1M_2 and ΔSM_2M_3 can be estimated as follows.

$$\begin{aligned} \Delta SM_1M_2 &= \sqrt{s_1(s_1 - R_1)(s_1 - R_2)(s_1 - l_1)} \\ &\qquad \textit{when}\, s_1 = \frac{1}{2}(R_1 + R_2 + l_1) \\ \Delta SM_2M_3 &= \sqrt{s_2(s_2 - R_2)(s_2 - R_3)(s_2 - l_2)} \\ &\qquad \textit{when}\, s_2 = \frac{1}{2}(R_2 + R_2 + l_1) \end{aligned} \tag{19.9}$$

When the equation of $l_1 = l_2 = l$ is applied, the sizes of the two triangles ΔSM_1M_2 and ΔSM_2M_3 become the same. As a result, the following equation can be established.

$$\begin{aligned} &\sqrt{s_1(s_1 - R_1)(s_1 - R_2)(s_1 - l)} \\ &\quad = \sqrt{s_2(s_2 - R_2)(s_2 - R_3)(s_2 - l)} \end{aligned} \tag{19.10}$$

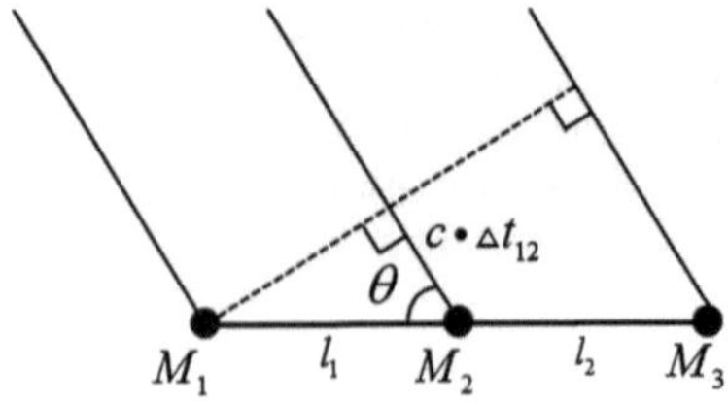

Fig. 19.2 Principles for the estimation of the sound-source angle in the two-dimensional space

By substituting Eq. (19.8) into the above equation and arranging the result, it is possible to estimate the distance d as follows.

$$d = \frac{l^2 + lc(\Delta t_{12} + \Delta t_{23}) - c^2\left(\Delta t_{12}^2 + \Delta t_{12}\Delta t_{23} + \Delta t_{23}^2\right)}{2c(\Delta t_{12} + \Delta t_{23})} \tag{19.11}$$

Estimation of the Angle in the Two-dimensional Space.

When the distances between the sound-source (S) and three microphones are sufficiently greater than the intervals of the microphones in the two-dimensional space, it is possible to assume that the sound waves reaching individual microphones are parallel. Figure 19.2 shows the sound waves reaching individual microphones in a parallel way.

Based on the microphone located in the center, the angle between a mobile robot and a sound-source can be regarded as θ. θ can be expressed as follows.

$$\theta = \cos^{-1}\left(\frac{c \cdot \Delta t_{12}}{l}\right) \tag{19.12}$$

19.3 Obstacle Avoidance by Using the Virtual Impedance

By using the virtual impedance method, it is possible to estimate the acceleration of the robot $\ddot{x}(s)$beyond the given trajectory by using spring and damping variables to model the relations between the current location of the mobile robot and the reference point and between the distance with an obstacle and the speed [13]. Figure 19.3 shows the case of applying the general impedance algorithm to the avoidance of obstacles by the mobile robot. In the process of modeling the distances and directions among the mobile robot, obstacles and the reference point, each category is shown as a point.

The virtual impedance algorithm generates a virtual kind of repulsive force based on the information regarding the distance between the robot and each obstacle and the related speed in order to make the robot avoid each obstacle. At the same time, there is a virtual kind of attraction force between the robot and the reference point which makes the robot move towards the reference point more effectively and smoothly. The equation of motion for the mobile robot, which can be

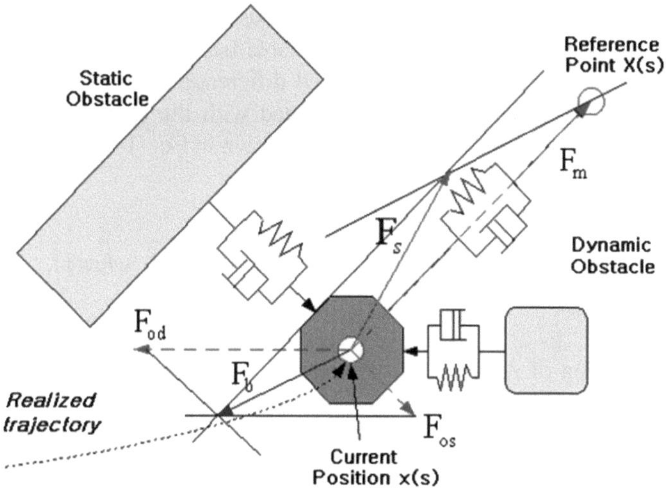

Fig. 19.3 Virtual impedance algorithm

obtained through the modeling process based on the virtual impedance method, can be expressed as follows.

$$\ddot{x}_s = \frac{1}{M_S}\left(F_m + \sum_{i=1}^{n_{os}} F_{os,i} + \sum_{i=1}^{n_{od}} F_{od,i}\right) \tag{19.13}$$

In this equation, n_{os} and n_{od} represent the numbers of the fixed obstacles and the dynamic ones respectively. M_S plays the role of the factor controlling the relative changing ratio of the robot trajectory. F_m shows the virtual attraction force between the current location of the robot x and the reference point X. F_m can be estimated as follows.

$$\begin{aligned} F_m &= K_r\left(X_{r,s}\right) + D_r\left(\dot{X}_{r,s}\right) \\ &= -K_r(x_s - X_r) - D_r\left(\dot{x}_s - \dot{X}_r\right) \end{aligned} \tag{19.14}$$

Here, K_r shows the spring variables applied between the current location of the robot x_s and the reference point X_r, while D_r shows the damping variables applied between the current speed of the robot $\dot{x}_s$and the moving speed $\dot{X}_r$of the reference point. In Eq. (19.13), $F_o$represents the virtual kind of repulsive force between the current location of the robot x_s and each obstacle. It can be estimated as shown in Eq. (19.15).

$$\begin{aligned} F_o &= K_o\left(X_{s,o}\right) + D_o\left(\dot{X}_{s,\ o}\right) \\ &= \sum_{i=0}^{n} \{K_o(x_s - x_{oi}) + D_o(\dot{x}_s - \dot{x}_{oi})\} \end{aligned} \tag{19.15}$$

Here, K_o and D_o represent the spring variable based on the distance between each obstacle and the robot and the damping variable based on the speed respectively. In the above equation, instead of the speed difference between the mobile robot and each obstacle, the damper can be defined with the change of the distance, inducing a new virtual impedance method as shown in Eq. (19.16).

$$\begin{aligned} F_o(s) &= K_{s,\,o}\left(X_{s,o}(s)\right) + D_{s,\,o}\left(\dot{X}_{s,\,o}(s)\right) \\ &= K_{s,\,o}(\rho_0 - \|d_i\|)\frac{d_i}{\|d_i\|} - D_{s,\,o}(\Delta(d_i))\frac{d_i}{\|d_i\|} \quad (when\|d_i\| < \rho_0) \end{aligned} \tag{19.16}$$

Here, the equation of $\Delta(d_i) = d_i[n] - d_i[n-1]$ is applied. ρ_0 shows the possible range for the estimation of the distance when the ultrasonic sensor is used. d_i shows the distance vector and can be estimated as $d_i = x_s - x_{oi}$. Since the locations of the mobile robot and obstacles are expressed as points, it is necessary to decide spring and damping variables in order to maintain a proper distance and avoid obstacles after considering individual volumes.

19.4 Experiment for the Recognition of the Sound-Source Location and Virtual Impedance Simulation

In this study, the experiment for the evaluation of the function performed by the algorithm of recognizing the sound-source location and the simulation of moving to the recognized location while avoiding obstacles were carried out. The experiment of recognizing the sound-source location was executed in the lobby of the building where the laboratory was located. In order to minimize the influence of the external noises, most of such factors as the existence of pedestrians or obstacles were removed. The simulation of the virtual impedance algorithm applied to the operation of the mobile robot was executed in each of the following cases; (1) when there was no obstacle, (2) when there was one obstacle on the expected route of the mobile robot, and (3) when there were more than two obstacles (in this paper, there were four obstacles).

19.4.1 Experiment for the Recognition of the Sound-Source Location

In order to evaluate the performance of the algorithm of recognizing the sound-source location, sound waves were generated with constant distances and angles. Speakers were used to create constant point sound-sources. Non-directional microphones were applied to receive sound waves. They were arranged with the intervals of 40 cm. They were made to face the front. During the experiment, the height of each speaker and the one of each microphone were made to be the same.

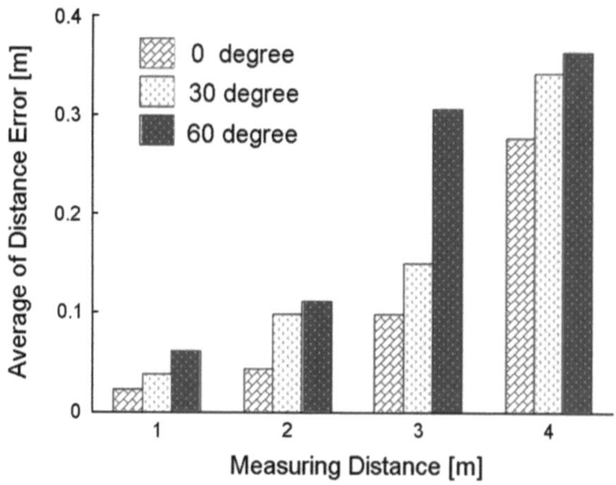

Fig. 19.4 Estimation error for the distance between the sound-source and the mobile robot

The experiment was carried out for the three cases with the angles of 0°, 30° and 60° between the front side of the mobile robot and the sound-source based on the microphone located in the center and the individual distances of 1, 2, 3 and 4 m with the sound-source. In order to improve the level of credibility for the experimental values, 30 experiments were carried out for each case. The results are shown in Fig. 19.4.

Figure 19.4 shows the average error for the data related to the estimation of the distance. According to the results of the experiment, as the estimated distance and the angle with the front side of the mobile robot became greater, the error also became greater. In particular, when the distance became more than 4 m, the error increased rapidly. In case of the distance being 3 m, the error increased based on the angle of 60°. Unlike the estimated results of the distance, the estimated results of the angle showed relatively constant errors in the entire area except for the field surrounding the angle of 0°.

19.4.2 Virtual Impedance Simulation

A simulation was carried out to apply the virtual impedance algorithm as the operational algorithm to move to the location estimated before. In order to calculate the impact vector between the mobile robot and obstacles in each simulation, individual locations were expressed in points. By considering the sizes of the mobile robot and obstacles, the repulsive force was created to make the robot avoid obstacles without a collision.

Case 1 When there is no obstacle—The virtual impedance algorithm is the method of operating the mobile robot by using the attraction force with the reference point and the repulsive force against obstacles. When there is no obstacle, the speed of the mobile robot is decided by only using the attraction force with the

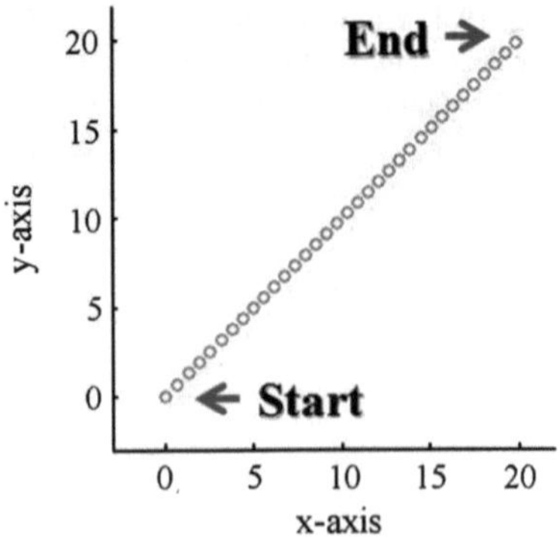

Fig. 19.5 Simulation results of Case 1

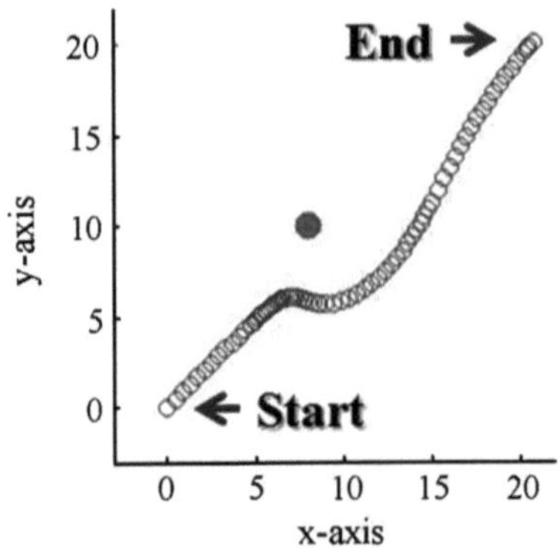

Fig. 19.6 Simulation results of Case 2

reference point. When the starting point is (0, 0) and the reference point is (20, 20), the expected trajectory can be drawn as follows (Fig. 19.5).

Case 2 When there is one obstacle on the expected moving route—By locating one obstacle near the expected route shown in case 1, the expected trajectory is drawn. The obstacle is fixed in the location of (8, 10) and shown in the form of a point.

In Fig. 19.6, when the mobile robot is faced with an obstacle while moving to the reference point, the robot gradually slows down as it approaches the obstacle until it changes its direction. After avoiding the obstacle to some extent, it increases its speed again and moves towards the reference point. By considering the sizes of the actual obstacle and the mobile robot, it is possible to see that the trajectory shown in the figure is enough for the avoidance of obstacles.

Case 3 When there are several obstacles on the expected moving route—A simulation is executed based on the assumption that there are several obstacles in random locations around the expected moving route between the starting point of (0, 0) and the reference point of (30, 30). The locations of the four obstacles are (5, 7) (13, 10) (19, 17), and (20, 27).

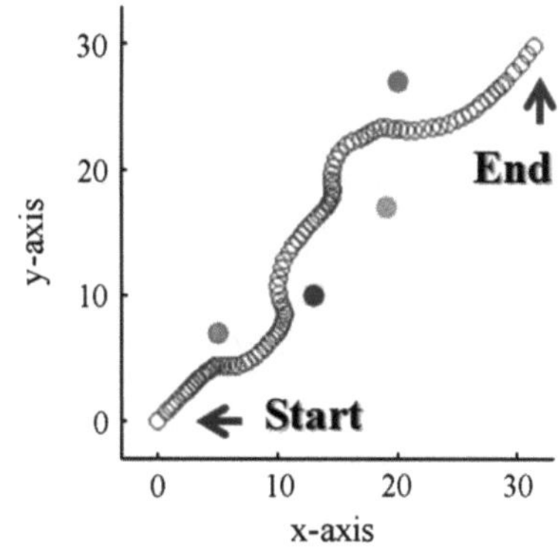

Fig. 19.7 Simulation results of Case 3

In Fig. 19.7, it is possible to see that the mobile robot gradually slows down as it approaches each obstacle in order to avoid it just like Case 2. The overall operational direction is towards the reference point. When there are obstacles around the operational route as shown in the figure, it is possible to see that the robot passes between the obstacles.

19.5 Conclusion

This paper focused on the case of the mobile robot avoiding obstacles in regard to the sound-source. An algorithm was suggested for estimating the positions of the randomly-located sound-sources by the mobile robot. Also, the algorithm for the avoidance of obstacles by the mobile robot when it approaches them while in motion in a certain environment equipped with obstacles was discussed. The operational direction of the mobile robot was set in the way of recognizing the locations of the obstacles while avoiding them and moving towards the reference point simultaneously by using the virtual impedance technique. The virtual impedance technique can be applied to both fixed and dynamic obstacles. However, in certain cases, it might not be able to avoid dynamic obstacles or have a flexible kind of trajectory. Also, when the mobile robot is located in the environment packed with obstacles and moves between them, it might collide with some obstacles without being able to avoid them or show abnormal changes for the speed. Therefore, it is necessary to study a flexible and safe operational method.

Acknowledgements "This research was supported by the MOTIE (The Ministry of Trade, Industry and Energy), Korea, under the Human Resources Development Program for Special Environment Navigation/Localization National Robotics Research Center support program supervised by the NIPA (National IT Industry Promotion Agency)." (H1502-13-1001). "This work was supported by the National Research Foundation of Korea (NRF) Grant funded by the Korean Government (MSIP) (NRF-2013R1A1A2021174)".

References

1. Kim Y-E, Hong S-A, Chung J-G (2009) Efficient implementation of IFFT and FFT for PHAT weighting speech source localization system. Inst Electron Eng Korea Signal Process 46(1):71–78 (WN.325)
2. Hu J-S, Yang C-H, Wang C-K (2008) Sound source localization by microphone array on a mobile robot using Eigen-structure based generalized cross correlation. In: IEEE international conference on advanced robotics and its social impacts, pp 1–6
3. Woo-han Y, Cheon-in O, Kyu-Dae B, Su-young J (2006) The impulse sound source tracking using Kalman filter and the cross-correlation. In: SICE-ICASE international joint conference, pp 317–320
4. Kwon B, Park Y, Park Y (2009) Multiple sound sources localization using the spatially mapped GCC functions. In: ICROS-SICE international joint conference, pp 1773–1776
5. Hao M, Lin Z, Hongmei H, Zhenyang W (2007) A novel sound localization method based on head related transfer function. In: The eighth international conference on electronic measurement and instruments, pp 4_428–4_432
6. Koch L, Adve R, Francis B (2007) Optimal beamforming with mobile robots. In: Conference on signals, systems and computers, pp 1652–1655
7. Huang J, Ohnishi N, Sugie N (1997) Building ears for robots: sound localization and separation. Artif Life Robot 1(4):157–163
8. Golfarelli M, Maio D, Rizzi S (2001) Correction of dead-reckoning errors in map building for mobile robots. IEEE Trans Robot Autom 17(1):37–47
9. Tsai C-C (1998) A localization system of a mobile robot by fusing dead-reckoning and ultrasonic measurements. IEEE Trans Instrum Meas 47(5)
10. Yi S-Y, Lee Suk-han, Hong Yeh-sun (1999) A real-time collision-free trajectory planning and control for a car like mobile robot. J Control Autom Syst Eng 5(1):105–114
11. Ge SS, Cui YJ (2002) Dynamic motion planning for mobile robots using potential field method. Auton Robots 13(3):207–222
12. Kitagawa L, Kobayashi T, Beppu T, Terashima K (2001) Semi-autonomous obstacle avoidance of omnidirectional wheelchair by joystick impedance control. In: Proceedings of IEEE/RSJ international conference on intelligent robots and systems, 200, vol 4, pp 2148–2153
13. Jin T-S (2009) Obstacle avoidance of mobile robot with virtual impedance. Korean Inst Intell Syst 19(4):451–456

Chapter 20
Object-Tracking Robot Using Data Combination of Ultrasonic Sensor and Servo Motor

Seung-Ik Hwang, Beom-Seok Seo and Jang-Myung Lee

Abstract In this paper, distance values of ultrasonic distance sensor and the rotation angle values of the servo motor are used as a way of tracking the moving object. Depending on the environment, it is needed to track the accurate path or the shortest distance by predicting them. So, to prove that the proposed tracking method which using ultrasonic distance sensor and servo motor as the input element is more accurate and affordable than tracking method using vision and LRF and multiple ultrasonic sensors as the input element, it is experimented to compare two tracking methods. Furthermore it is proved that tracking is more accurate by applying Intermediate filter.

Keywords Object-tracking robot • Ultrasonic sensor • Servo motor • Rotation angle • Mobile robot

20.1 Introduction

Worldwide, according to industrial development and improvement of standard of living, research on robots for convenience has actively been done. There are some examples cleaning robots, service robots, and industrial robots. They show robots ties closely with people for a variety of purposes in real life or industrial. Basically, in order to drive the robot, the input element and the output element

S.-I. Hwang · B.-S. Seo · J.-M. Lee (✉)
Department of Electrical Engineering, Pusan National University, Busan, South Korea
e-mail: jmlee@pusan.ac.kr

S.-I. Hwang
e-mail: seungik7379@pusan.ac.kr

B.-S. Seo
e-mail: beomseok7379@pusan.ac.kr

R. Tutsch et al. (eds.), *Progress in Optomechatronic Technologies*,
Lecture Notes in Electrical Engineering 306, DOI: 10.1007/978-3-319-05711-8_20,

should be present. Several sensors can be examples of input elements. There are several kinds of sensors, distance measurement sensors, location-aware sensors, vision, pressure sensors, microphones, etc. Robot can be driven by calculating or by analyzing input data received from these sensors. Also driving element means the output element. The output element is electric device such as motor or hydraulic devices, monitors, speakers. Along with the development of the intelligent robot technology, many methods have been proposed for the interaction of people and robots. Mobile robot moving along the target is one of the ways in interaction of humans and robots. Also applying robot to the real world is becoming more important. In order to implement tracking object, calculate rotate error and straightness error between robot and objects by combining ultrasonic distance sensor to servo motor and by using the distance data measured by ultrasonic distance sensor and the angle of the servo motor data. Thus, by using the calculated error and by controlling two DC motors Installed on the robot, let robot track moving object in maintaining a certain distance. In addition to the ultrasonic distance sensor, there are vision and laser finder it is mainly used at a tracking robot. Although vision through image processing is more accurate than other sensors to achieve environment information, it needs a High specification of the system to process large data. As a result, it costs high. Also laser finder is affected by many obstacles surrounding environment, that's why it is difficult to obtain reliable data. Also its uncertainty is so enlarged that tracking is likely to fail. In contrast, ultrasonic distance sensor is less impacted on the surrounding environment and has advantage of high accuracy of the data at low cost comparing with vision and laser finder [1].

There are various methods of tracking object through using ultrasonic distance sensor, but among them, frequently used method is to pace number of ultrasonic distance sensors in different angles to tract direction in the sensor of being recognized. However, there are disadvantage of low accuracy. As direction of object is changed, ultrasonic angle of reflection is changed or sound wave from each sensor can be input to different sensor. This phenomenon shows inaccurate value is recognized [2].

To complement previously mentioned method and to track object in maintaining a certain distance from the robot to objects, straightness errors and rotation error generated by distance value of ultrasonic distance sensor, the value of the angle of the servo motor are calculated. Also the speed of the robot is controlled by using proportional control. Finally two experiments are discussed by comparing one inserting a filter in the middle of the ultrasonic distance sensor with the other not inserting. The filed this method is used is auto golf cart [3] or multiple robots moving in the formation at a constant distance [4]. In this paper, it proposes input elements which are ultrasonic distance sensor, description for the features, performance of servo motor and how to take advantage of servo motor in Sect 20.2. Also Wheel-drive robot and control techniques for moving object tracking using rotate are described in Sect. 20.3 and performance it is suggested by experiment of moving object tracking by using graph is described in Sect. 20.4. Finally it concludes this paper in Sect. 20.5.

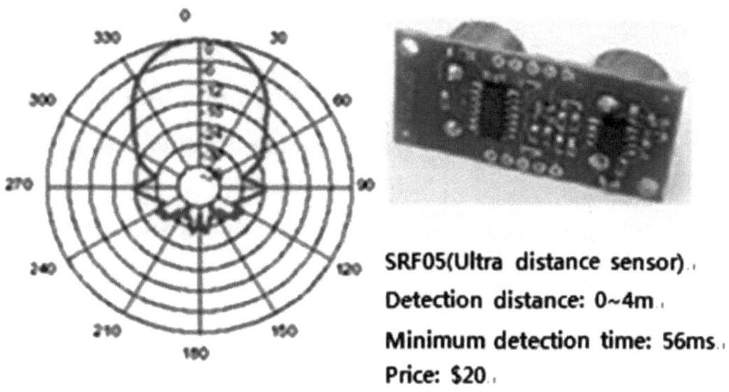

Fig. 20.1 Performance of ultrasonic sensors were used in the experiment

20.2 Explanation and Using Method of Distance Sensor Servo Motor

20.2.1 Ultrasonic Distance Sensor

Ultrasonic distance sensor is device using a frequency band of sound waves humans can't hear and it forms a one set of transmitter and receiver. There are two type of sensor. One is the reflective type that Transmitter and receiver lean towards each other in the same direction the other is the opposite type that the transmitter and receiver face each other. In this paper, reflective type is used. More performance is shown in Fig. 20.1.

The operation Method of the sensor like as shown in the Fig. 20.1. Distance can be calculated by the round-trip time from ultrasonic sensor to obstacle. The ultrasonic distance sensor varies in performance but low-cost sensors were used. Its measurement error is +5 cm, output range is 0–4 m, measuring angles is up to 25° and the minimum detection time is 56 ms.

20.2.2 Servo Motor

Basically servo motor is driven by receiving a PWM signal input, as shown in Fig. 20.2. Motor can indicate the direction continuously depending on the input pulse width. Servo motor used in the experiment can move up to 120° and its input voltage is 4.8–6 V. The round-trip time in motor is from minimum 0.64 s to maximum 0.84 s.

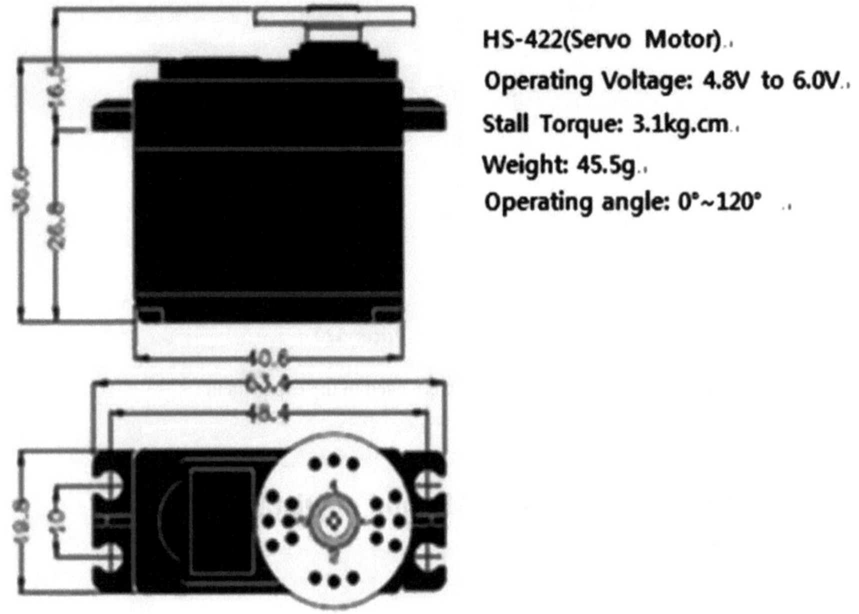

Fig. 20.2 Performance of servo motor were used in the experiment

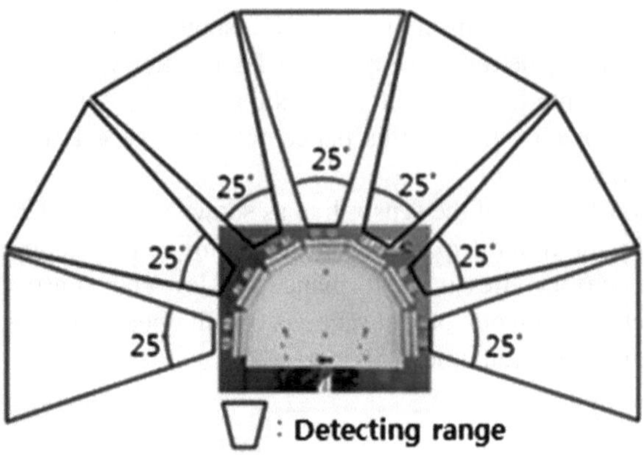

Fig. 20.3 Range and deployment method using a plurality of ultrasonic distance sensor

20.2.3 Taking Advantage of the Ultrasonic Distance Sensor and Servo Motor

In order to track object, it is needed to determine the direction and distance of an object. Mobile robot can move using only one ultrasonic sensor but there are one data of distance not both information of direction and distance. The direction of the object can be known by Multiple Ultrasonic distance sensors installed in several directions as shown in Fig. 20.3. Seven sensors of detection angle of

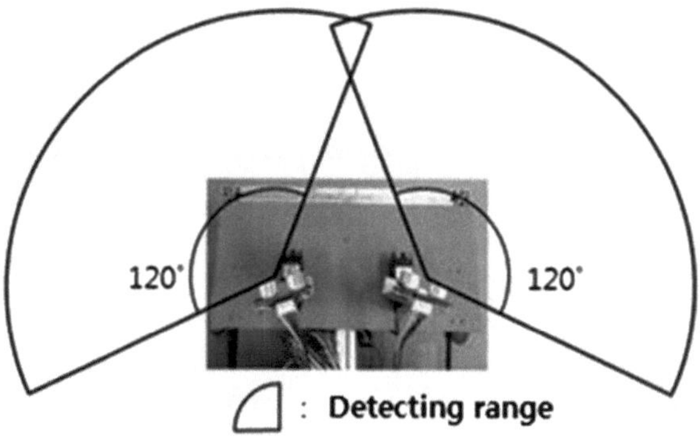

Fig. 20.4 Placement method and range that combines ultrasonic distance sensor and servo motor

25° are installed to detect the direction of more than 180°. Also, information of distance and direction can be known by combining ultrasonic distance sensor with servo motor that can designate direction through method suggested in this paper. Ultrasonic distance sensor can detect 120° range and two sensors are installed in robot as shown in Fig. 20.4.

20.3 Tracking Control of Mobile Robot

20.3.1 Configuration and the Equations of Motion for Mobile Robot

Configuration of mobile robot inputs is consisted of two sensors combined ultrasonic distance sensor with servo motor set up on the top of mobile robots using a fixed device as shown in Fig. 20.5. The detection range of 120° is respectively servo motor's reference value. Then, it is possible to minimize the Straightness errors when object is detected by two sensors. The reason why two combined sensors set up is that the detection of the round-trip time can be reduced half by using two servo motors instead of one. In addition, this configuration has the advantage of detecting wide area better than fixed ultrasonic sensor for navigate the specified direction. So, it has the same efficiency using multiple ultrasonic sensors.

Likewise, output element is two dc motor as wheel drive. Figure 20.6 shows the position of the robot according to the movement of wheel-drive mobile robot.

Position can be represented by Eq. (20.1).

$$P(t) = [x(t), y(t), \theta(t)] \tag{20.1}$$

$x(t)$, $y(t)$ are coordinates of the position about robot over time respectively and $\theta(t)$ is the angle between the X-axis and the front of the robot. Straight speed $V_S(t)$ of the robot and the rotational speed $V_R(t)$ are shown in Eq. (20.2). r is the radius of

Fig. 20.5 Configuration of a mobile robot equipped with ultrasonic sensors and servo motor

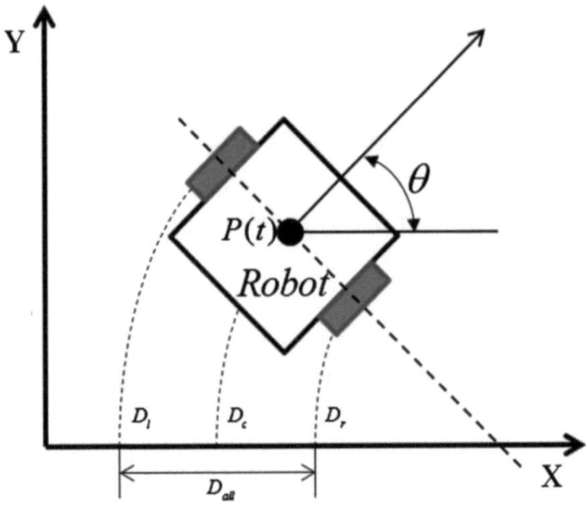

Fig. 20.6 Movement and position of the wheel-drive mobile robot

the wheel, ω_l is the angular velocity of the left wheel and ω_r is the angular velocity of the right wheel.

$$\begin{aligned} V_S &= \tfrac{r}{2}(\omega_l + \omega_r) \\ V_R &= \tfrac{r}{2}(\omega_l - \omega_r) \end{aligned} \tag{20.2}$$

Position and change of direction according to the movement for mobile robot are separated by two cases. One case is robot moves in no rotation, the other case is robot moves in rotation. The former case can be represented by Eq. (20.3), the latter case can be represented by Eq. (20.4).

- In case of not having rotational motion.

$$\begin{aligned} \Delta x &= V_S \Delta t \cos\theta(t) \\ \Delta y &= V_S \Delta t \sin\theta(t) \\ \Delta\theta &= 0 \end{aligned} \tag{20.3}$$

- In case of having rotational motion.

$$\begin{aligned} \Delta x &= \tfrac{V_S}{V_R}\{\sin[\theta(t) + \Delta\theta] - \sin\theta(t)\} \\ \Delta y &= -\tfrac{V_S}{V_R}\{\cos[\theta(t) + \Delta\theta] - \cos\theta(t)\} \\ \Delta\theta &= V_R \Delta t \end{aligned} \tag{20.4}$$

Therefore, consecutive location and position of the mobile robot in an absolute coordinate system can be represented by Eq. (20.5).

$$\begin{aligned} x(t+1) &= x(t) + \Delta x \\ y(t+1) &= y(t) + \Delta y \\ z(t+1) &= z(t) + \Delta z \end{aligned} \tag{20.5}$$

20.3.2 Tracking Control of Robot for Moving Objects

The configuration of mobile robot for moving object is as follows. In here, it assumed that object exists only or closer than other objects in experimental environment. Then, mobile robot is driven to maintain a certain distance with target for tracking. Direction angle of servo motor ϕ_d, distance of the target value S_d are given information. Distance value S_d can be found by using ultrasonic distance sensor and each direction angle ϕ_d can be found in real-time by using the angle of rotation of the servo motor. Figure 20.7 shows the state for the actual moving object tracking and then, represented by a symbol. The moving speed and the rotational speed are necessary for mobile robot to drive. So, two speeds should be found. Using the distance and direction obtained from the sensor, velocity it can make S converge to S_d and ϕ converge to ϕ_d should be found. Such expression is represented by Eq. (20.6).

$$l = s \cdot \sin(\phi - \omega) \quad \therefore \frac{l}{s} = \sin(\phi - \omega) \tag{20.6}$$

Therefore, ω can be represented by the Eq. (20.7).

$$\omega = \phi - \sin^{-1}\left(\frac{l}{s}\right) \tag{20.7}$$

l_t can be derived like as Eq. (20.8) in Fig. 20.7.

$$l_t = s \cdot \cos(\phi - \omega) - s_d \cdot \cos\phi_d \tag{20.8}$$

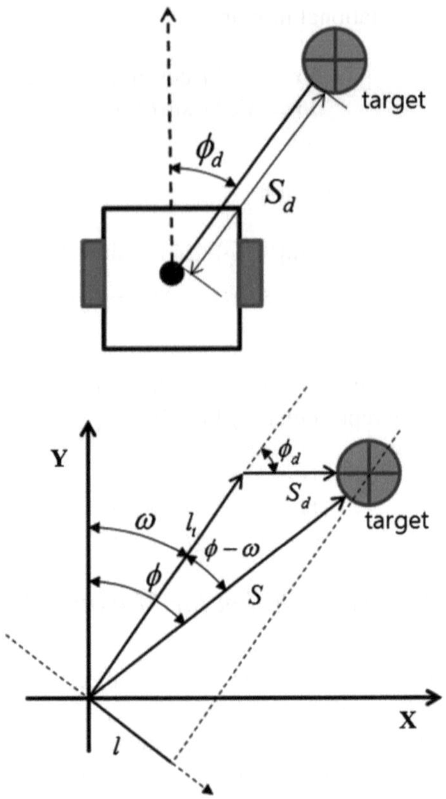

Fig. 20.7 Movement and position of the moving object tracking

Based on the calculated expression, PWM of DC motor it is equipped in both sides of the mobile robot was controlled proportionally to converge it because l_t and ω are zero.

20.4 Experiment and Comparison

20.4.1 Experimental Methods

Experiment was done by producing comparison of target to prove method suggested was effective. Comparison of target is mobile robot with multiple ultrasonic distance sensor, which uses 7 same ultrasonic distance sensor and it has Probe range of 180°. The experimental conditions include different input elements but same equipment and same place. Similarly configuration of robot's body, velocity about straight error and rotation error are same experimental conditions. Experiment and comparison are done through 90° curve driving and S-shaped driving. As the result, it can be represented from the sensor information. Also two cases are compared with applying intermediate filter to Ultrasonic distance sensor and not applying it.

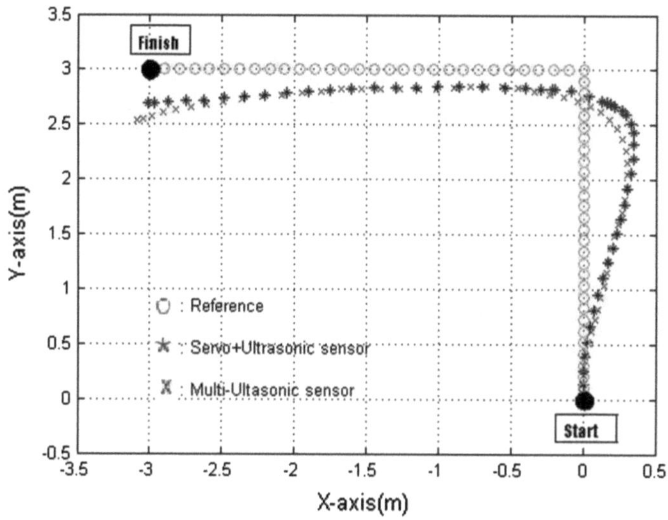

Fig. 20.8 90° traveling of the robot movement trajectory graph (not applying filter)

Table 20.1 Experimental error of 90° traveling (not applying filter)

	Servo motor		Multiple sensors	
	X-axis	Y-axis	X-axis	Y-axis
Average error (cm)	16.5	13.3	17.5	14.8
Maximum error (cm)	32.1		42.3	

20.4.2 The Experiment Not Applying the Intermediate Filters

- Case 1: 90° traveling
- Case 2: S-shaped traveling

From Figs. 20.8, 20.9, 20.10 and 20.11 for each of the symbols, O shape of the green is a reference, * shape of the blue is trajectory of the robot using an ultrasonic distance sensor and two servo motors. And X shape of the red is trajectory of a mobile robot using multiple ultrasonic distance sensors. These three elements Integrate into one graph to compare them. Experimental result for first 90° driving is shown in Fig. 20.8 and Table 20.1. And there was no significant difference when comparing the overall path. Overall, the average error was about 1 cm and the maximum error was about 10 cm. Second S-curve driving was differences in movement. At the first starting point, mobile robot with multi-ultrasonic sensor move across the inside of the curve without following the reference. On the other hand, mobile robot combined servo motor and ultrasonic sensor moves outside slightly beyond the reference. The reason why this migration path was shown in graph is that multi-ultrasonic sensor detects several directions but servo ultrasonic sensor can't detect interval during rotation. It causes rotation time error in using

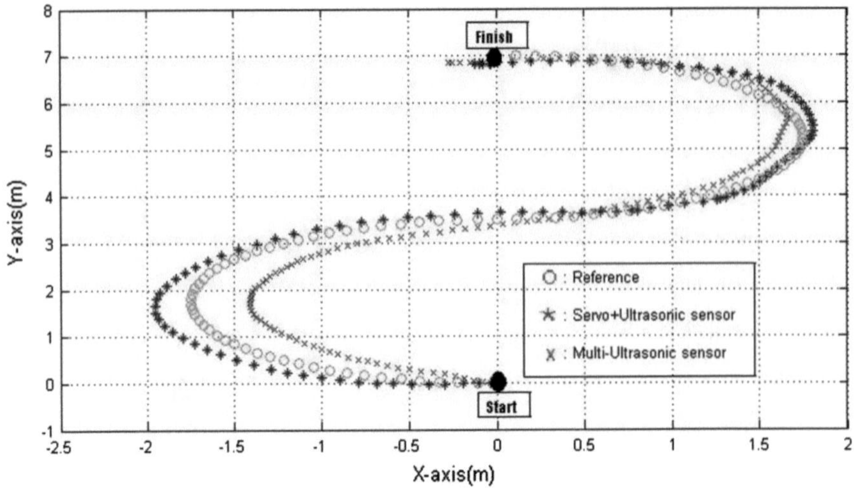

Fig. 20.9 S-shaped traveling of the robot movement trajectory traveling (not applying filter)

Table 20.2 Experimental error of the S-shaped traveling (not applying filter)

	Servo motor		Multiple sensors	
	X-axis	Y-axis	X-axis	Y-axis
Average error (cm)	13.8	17.2	20.7	16.9
Maximum error (cm)	23.5		33.9	

servo ultrasonic sensor. Also, errors in the two paths, servo motor's average error is smaller than multi-ultrasonic sensor's error about 7 cm to the X-axis but bigger than its error about 0.3 cm to the Y-axis. The maximum error is smaller than its error about 10 cm. Finally, the experiments are carried out by applying the intermediate filter to compensate for the path of movement (Table 20.2).

20.4.3 The Experiment Which is Applied the Intermediate Filters

- Case 1: 90° Traveling

Experimental result for the first 90° travelling applied in intermediate filter is shown in Fig. 20.10 and Table 20.3.

As compared with average error, servo motor's average error is smaller than multi-ultrasonic sensor's error about 3 cm to X-axis and bigger than its error about

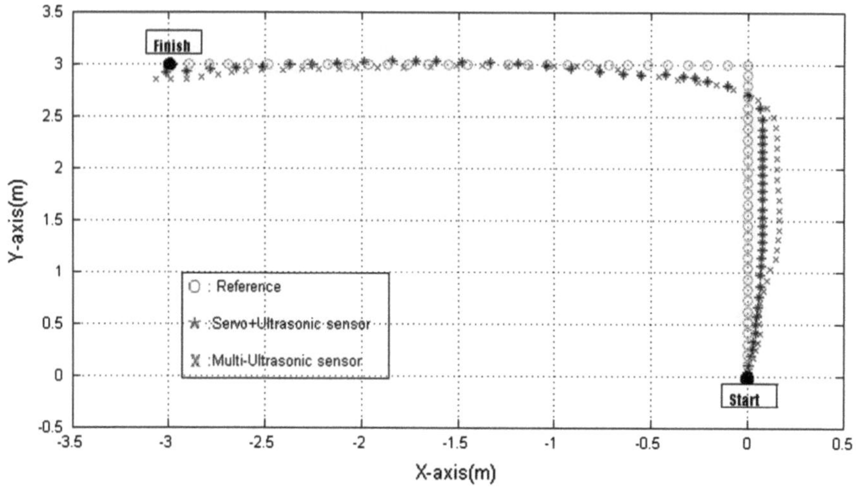

Fig. 20.10 90° traveling of the robot movement trajectory graph (applying filter)

Table 20.3 Experimental error of 90° traveling (applying filter)

	Servo motor		Multiple sensors	
	X-axis	Y-axis	X-axis	Y-axis
Average error (cm)	10.5	10.3	13.8	9.2
Maximum error (cm)	17.3		19.6	

1 cm to Y-axis. The maximum error is smaller than its error about 2.3 cm. Despite straight driving is not a high different, this result tells servo motor is more effective when mobile robot track the moving object at the point of total path.

- Case 2: S-shaped traveling

Experimental result for S-shaped travelling applied in intermediate filter is shown in the Fig. 20.11 and Table 20.4. Servo motor's average error is smaller than multi-ultrasonic sensor's error about 8 cm to X-axial and smaller than its error about 1 cm to Y-axial. The maximum error is notably smaller than its error about 16 cm. This results show method of tracking using servomotor depending on the ratio of the angle of rotation is more effective because it's maximum error is smaller than method of tracking using multi-sensor's maximum error about 16 cm. This shows that accurate tracking. Comparing filter applied mobile robot with filter not applied mobile robot, there is no significant difference in multi-ultrasonic sensor but servo ultrasonic sensor is significantly improved as filter apply mobile robot.

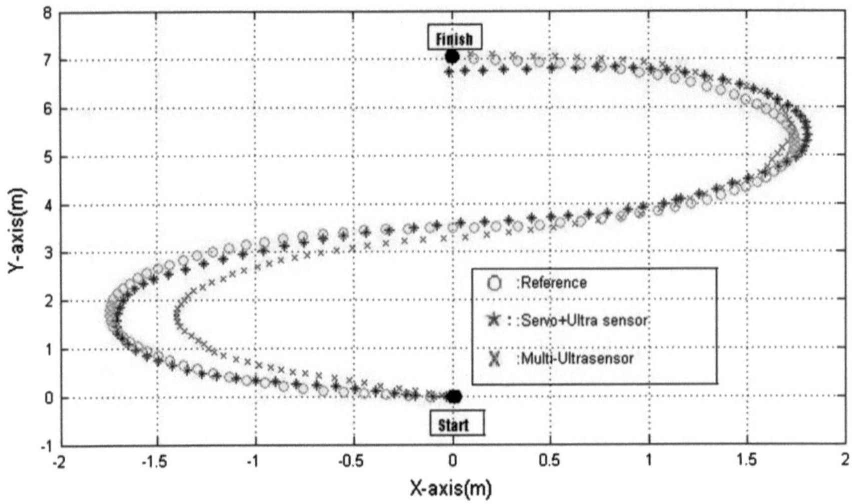

Fig. 20.11 S-shaped traveling of the robot movement trajectory traveling (applying filter)

Table 20.4 Experimental error of the S-shaped traveling (applying filter)

	Servo motor		Multiple sensors	
	X axial	Y axial	X axial	Y axial
Average error (cm)	10.3	13.8	18.5	14.5
Maximum error (cm)	18.6		34.7	

20.5 Conclusion

In this paper, moving object tracking method using ultrasonic distance sensor and servo motor is proposed. Features of the proposed method are as follows. Input element for tracking is ultrasonic distance sensor.

- Data size of ultrasonic distance sensor is smaller than camera or LRF sensor and process time is faster than them. Also, it suited to real-time target tracking.
- Ultrasonic distance sensor and servo motor are used to gather the data about distance and direction in tracking obstacle.
- Method in this paper is more efficient for tracking than using method of seven ultrasonic distance sensor.
- In case that the intermediate filter is applied, the performance of the proposed method is improved.

It is explained that the proposed mobile robot for tracking moving objects using a low-cost ultrasonic distance sensor and servo motor can track the moving objects efficiently. By placing a plurality of ultrasonic sensors, the interference of the ultrasonic sensor which having a disadvantage of tracking can be solved. In addition, it shows high-efficiency, accurate tracking at a low-cost.

Acknowledgements "This research was supported by the MOTIE (The Ministry of Trade, Industry and Energy), Korea, under the Human Resources Development Program for Special Environment Navigation/Localization National Robotics Research Center support program supervised by the NIPA (National IT Industry Promotion Agency)." (H1502-13-1001). "This research was financially supported by the Ministry Of Trade, Industry and Energy (MOTIE), Korea Institute for Advancement of Technology (KIAT) and DongNam Institute For Regional Program Evaluation (IRPE) through the Leading Industry Development for Economic Region."

References

1. Han S-B, Lee S-G, Lim J-K, Kim D-Y, Kim J-H (2006) The study on map building algorithm and traveling algorithm of the mobile robot with the ultrasonic sensor. Korean Soc Mar Eng 0:1–292
2. Lee S-J, Lim J-H, Cho D-W (2010) A complete feature map building method of sonar sensors for mobile robots. J Korea Soc Precis Eng 27(1):64–75
3. http://www.golfingtrolley.com/
4. Park S-S, Kim K-J, Ko NY, Seo DJ (2007) Position estimation of a mobile robot using electronic compass and encoder. Korea Inst Inf Technol 5(4):1–256 (12)

MIX
Papier aus verantwortungsvollen Quellen
Paper from responsible sources
FSC® C105338

If you have any concerns about our products,
you can contact us on
ProductSafety@springernature.com

In case Publisher is established outside the EU,
the EU authorized representative is:
Springer Nature Customer Service Center GmbH
Europaplatz 3, 69115 Heidelberg, Germany

Printed by Libri Plureos GmbH
in Hamburg, Germany